CONNECTING ARDUINO

PROGRAMMING AND NETWORKING WITH THE ETHERNET SHIELD

BOB HAMMELL

Connecting Arduino: Programming and Networking with the Ethernet Shield

Published in the United States of America by Bob Hammell.

ISBN-10 (Print): 1-500-74567-7
ISBN-13 (Print): 978-1-500-74567-7
ISBN-13 (ePub): 978-1-312-41034-3

Any source code or supplementary materials referenced by the author in this text are available for readers at www.connectingarduino.com.

Table of Contents

Preface

At this point, the Arduino hardly needs any introduction. It's become a force of nature – inspiring, in its short lifetime, millions of people from all walks of life and with varying levels of prior experience in electronics and computer programming. There's much you can do with this flexible development platform, and so much amazing work has already been done. But where things really get interesting, really get useful, is when you make projects that talk to each other and to the rest of the world.

That hobbyist and beginner electronics hackers and "makers" can create standalone devices which communicate with other machines on the local network and across the Internet, and using the same Internet protocols as used by desktop PCs, servers, and mobile devices, is certainly not insignificant.

Despite the emergence of new development boards, shields, and modules, the Ethernet Shield remains a popular choice for Arduino projects. And it's easy to see why – the section of this book that covers getting the shield up and running is very thin. Unfortunately, making full use of this ingenious device is a little more difficult than the first steps suggest...And that brings me neatly to the subject of protocols and the reason why I wrote this book.

What's in this Book?

Internet and network communication is made up of many layers – starting with low-level protocols and techniques used to handle communication with hardware devices, such as network cards, modems, and Wi-Fi dongles. On top of this layer, the Internet protocol (IP) is responsible for the delivery of message fragments (or packets) to the intended recipient. Then, running over IP, you have the transport layer where transmission control protocol (TCP) adds error-checking and streaming capabilities. The application layer consists of protocols such as hypertext transfer protocol (HTTP), domain name system (DNS), and simple mail transfer protocol (SMTP). These *application protocols* define how data is encoded and exchanged for a specific purpose. You could also say that web-based application programming interfaces (APIs) and web services which run over HTTP add a fourth layer to this system.

The Ethernet Shield, in partnership with the Ethernet library that comes with the Arduino IDE, does an excellent job of encapsulating the complexities of TCP/IP and talking to the Wiznet W5100 integrated circuit on the shield. But its help only goes as far as the transport layer; you're on your own when it comes to HTTP and the application layer. At first, working with these protocols seems a daunting task – one that appears that only accomplished and experienced programmers have the skills or knowledge to attempt – and this suggestion is reinforced by the relatively small number of examples and guides that really try to explain the details of application protocols to Arduino programmers. So if you learn only one thing from this book then I hope it is this: working with application protocols is nothing to be afraid of.

I've written Connecting Arduino to show you, in quite a lot of detail, how to use application protocols in your Arduino sketches and get the most out of the Ethernet Shield. The majority of the information is organized into eight "projects" – and I use that term loosely. The goal was not to give you a recipe book, or a collection of plans for Arduino projects. Instead, I want to walk you through the background information, library classes and methods, and programming techniques that you can use in your own projects. But, critically, I wanted to give each item enough contextual information so that it's easy for you to see its relevance to the task at hand. The chapters divide into themes, and each project builds on the knowledge and information presented in the previous project. As such, you may find it beneficial to read through the book in order, even if you do not actually build and complete each project.

My projects might seem basic, but the ones you develop yourself afterwards will be much more interesting. And I hope you let me know about the cool things you build – or better still, have the devices contact me themselves.

Who Should Read this Book?

Unfortunately, I can't teach you everything about the Arduino in the space of one book. There's too much about electronics, C, and programming in general to cover. I have to assume that you're already competent at programming the Arduino and building simple circuits. With this said, if you can connect a light-emitting diode (LED) to the Arduino, through an appropriate resistor, and write a sketch that turns the LED on and sends a message to the Arduino's serial port then you'll easily understand 90% of the code and circuitry in this book.

For some of the projects, basic familiarity with hypertext mark-up language (HTML) would be useful. Building webpages and web-based user interfaces is definitely a skill, but it is one that you can learn as you go, and there is no shortage of excellent tutorials available online to help you.

Online Resources

ConnectingArduino.com is the companion website for this book; you can contact me there if there's anything I can help you with or if you want to show off your work. I've also put all of the project sketches up there so that you can download them, instead of typing them in. It'll be worth your while to visit the site regularly – any news, updates, and addendums will be posted there first.

To the best of my ability, I have verified the accuracy of all of the information in this book, and tried to ensure that the code samples are robust enough for you to use (while not being so full of optimized programming code and error-checking as to make the code difficult to understand). However, things change and mistakes do happen. You can help me to improve future editions, for the benefit of other Arduino enthusiasts, by contacting me at the website if you find any errors, inaccuracies, or places where information is confusing.

Conventions Used in this Book

The following table describes the text conventions used in this book.

Convention	Meaning
Italic	Text that appears in italics refers to file names, variable and function names, or other code that exists in the project sketch or Arduino libraries. Within the context of giving instruction, italic text should be typed exactly as shown.
Bold	Within the context of giving instruction, items in bold text are user interface elements, such as key strokes, menu items, or button labels. In other contexts, words may be emboldened for emphasis.
`Monospace font`	A monospace font is used for Arduino C, Processing, JavaScript, and HTML code that should be typed in your project.
Colored text	Items shown with colored text are links to other pages in this book.

Getting Started

The Arduino Ethernet Shield is an additional circuit board that fits on top of your Arduino. It extends the Arduino's capabilities with circuitry to connect to a network router, using a commonly-available RJ45 Ethernet cable. Your Arduino projects can communicate with the world through this connection – everything from fetching information from the Internet and displaying it on a liquid crystal display (LCD), to providing publically-accessible, web-based tools that can control motors and other hardware.

More than just a hardware device that can consume content and accept messages, the Ethernet Shield is your entry point into building things for the Internet of Things – devices that take an active role in talking to humans and other machines over Internet protocols.

In This Chapter

Connecting the Shield

The Arduino Ethernet Shield R3 mounts on top of Arduino devices using long wire-wrapped headers that extend through the shield and into the headers of the Arduino below. It only fits in one direction.

To connect the shield to an Arduino Uno R3 or Arduino Leonardo:

1. Disconnect the Arduino from all power sources, and remove any wires connected to it.

2. Line up the shield's headers with those of the Arduino.

3. Apply gentle pressure until the shield slots securely into place.

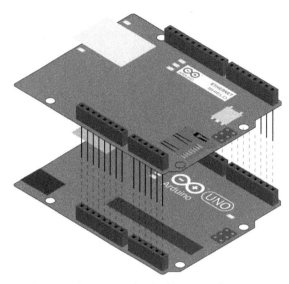

Figure 1. Connecting the shield to an Arduino Uno

To connect the shield to an Arduino Mega 2560:

1. Disconnect the Arduino from any source of power and remove any wires connected to it.

2. Line up the shield's headers with those of the Arduino. The shield slots into the two left-most groups of headers – up to RX0 on the top row and A5 on the bottom row.

3. Apply gentle pressure until the shield slots securely into place.

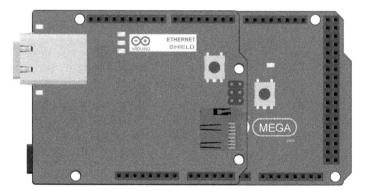

Figure 2. Connecting the shield to an Arduino Mega 2560

The Ethernet Shield R3 can also be used with earlier Uno devices and the Duemilanove. However, when using older Arduinos, four of the shield's header pins are left unconnected. You must ensure that none of these pins are allowed to make contact with any of the Arduino's components, or each other.

Suitable options for this are:

- Wrap the two left-most pins on the top row, and the two left-most pins on the bottom row, in insulating tape.
- Bend the two left-most pins on the top row, and the two left-most pins on the bottom row, away from contact with the Arduino.

Once the shield is fitted securely on the Arduino, you can reconnect the power.

> **Caution: It is usually safe to connect and disconnect cables and wires from the Ethernet shield while the Arduino is connected to its power supply. However, to avoid any accidental damage to electronic components, it is preferable to disconnect the power before doing so.**

The connectors and key components of the Arduino Ethernet Shield R3 are shown below:

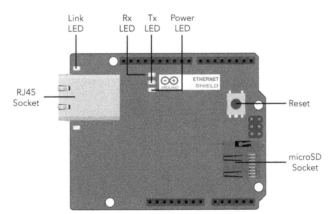

Figure 3. The Arduino Ethernet Shield

It is possible to stack other shields on top of the Ethernet Shield, and to use most of the Arduino's pins as usual. However, the Arduino talks to the Ethernet Shield over SPI and when actually using the Ethernet Shield, the following pins are unavailable for any other purpose:

Arduino Uno Pin	Arduino Mega Pin	Function
D4	D4	SS – when using the SD card.

Arduino Uno Pin	Arduino Mega Pin	Function
D10	D10	SS – when using the Ethernet Shield's SPI interface.
D11	D50	MOSI
D12	D51	MISO
D13	D52	SCK
–	D53	Not used. But must not be set to an input.

There are several methods of cabling the Ethernet Shield to your network. This choice makes no difference to how Arduino sketches are programmed, and you should simply choose the one that is most convenient for you and your workspace.

Connecting the Ethernet Shield to a Router

To connect the Ethernet shield to a router:

1. Plug one end of a CAT5 or CAT6 Ethernet cable with RJ45 connectors into the socket on the Ethernet Shield.
2. Plug the other end of the cable into an available Ethernet port on your router.
3. Plug the Arduino into a suitable power supply (if it is not connected already).

Using PowerLine Adapters

Unless you have a very long Ethernet cable, it may not be convenient to cable your Ethernet Shield directly to your router.

PowerLine adapters are devices that plug into electrical sockets and send computer signals along the power lines in your home. They are sold in pairs: one is to be positioned near to the router and one is to be used where needed.

These adapters require no configuration and work well with the Arduino Ethernet Shield.

Connecting the Ethernet Shield through a Bridged Connection

Using a standard Ethernet cable (or a crossover cable, if you have a really old PC), you can connect the Ethernet Shield to your PC and share its network connection.

1. Plug one end of a CAT5 or CAT6 Ethernet cable with RJ45 connectors into the socket on the Ethernet Shield.

2. Plug the other end of the cable into a free Ethernet port on your PC.

3. Plug the Arduino into a suitable power supply (if it is not connected already).

For the shield to be able to connect to the network, you must "bridge" the connection that your PC uses to connect to the network with the connection that is made to the Ethernet Shield. The process for doing this may be different depending on the operating system that you are running.

On Windows 8/7/Vista/XP:

1. Press the **Windows logo key + R**.

2. Type *ncpa.cpl* and press **Enter**.

3. Hold down the **Ctrl** key and click both the network connection that is used by your PC to connect to the network, and the network connection that is connected to the Ethernet Shield.

4. Right-click one of the selected connections, then click **Bridge Connections**.

On Mac OS X you can share your Mac's Wi-Fi connection with the Arduino Ethernet Shield:

1. On the **Apple** menu, click **System Preferences**, and then click **Sharing**.

2. On the sidebar, click **Internet Sharing**[1], and choose the Internet connection you want to share from the "Share your connection from" menu.

3. Select the checkbox labelled "Built-in Ethernet".

4. Click **Start**.

If your PC's operating system warns you that it has detected an IP address conflict, you may have to connect either your PC or the Arduino to the network using a static IP address.

---

1. In early versions of Mac OS X, click the **Internet** tab.

Establishing a Network Connection

The RJ45 socket on the Ethernet Shield contains two light-emitting diodes (LEDs). The left LED is the link indicator and glows or blinks green if a successful link has been made to the router. You can also find this same indicator as a surface-mounted LED just above the RJ45 socket on the shield. If the link LED is not lit at all, check your connections and replace the Ethernet cable if possible. In some circumstances, it may also be useful to restart your router.

A green LED does not mean that the Arduino is now connected to the network, only that communication between the shield and the router is working. To actually make a full connection to the network, the Arduino must be programmed with a sketch that uses the Ethernet library to set several configuration options.

In this section you will see how to create a basic sketch that connects to your network over dynamic host configuration protocol (DHCP), and how to test that your Arduino is properly connected.

Starting a New Sketch

In the Arduino integrated development environment (IDE), start a new sketch.

The Arduino talks to the Ethernet Shield over serial peripheral interface (SPI), and so to work with the shield you must include both the Ethernet and SPI libraries in your project. To do this, add the following two lines to the top of the sketch:

```
#include <SPI.h>
#include <Ethernet.h>
```

Specifying a MAC Address

The next piece of information that is usually defined in the sketch is the media access control (MAC) address. A MAC address is a 48-bit number (usually expressed as 6 bytes) that uniquely identifies a device on a local area network.

These numbers are usually built-in to the device and never change. You can generally find the MAC address for your Arduino Ethernet Shield printed on a sticker on the underside of the shield, or on the box that the shield came in. But if you do not have one, it is usually fine to make up six random numbers from 0 through 255. It is highly unlikely that you will randomly choose an address that is currently being used by another device on your network

MAC addresses can also be bought from the IEEE Registration Authority, or you can buy a read-only memory (ROM) chip that is pre-programmed with a unique address. Microchip Technology Inc. and Maxim Integrated Inc. offer a range of low-cost chips of this type.

The MAC address (either randomly generated or purchased) is typically included in the Arduino sketch as a global array of bytes:

```
byte mac[] = { 0x00, 0xC3, 0xA2, 0xE6, 0x3D, 0x57 };
```

As it is unlikely that you will ever need to modify this address while the sketch is running, you can also define the MAC address using a constant array:

```
const byte mac[] = { 0x00, 0xC3, 0xA2, 0xE6, 0x3D, 0x57 };
```

Completing the Sketch

The full Arduino sketch is shown below. This example connects to the network using DHCP, during which the router assigns connection properties to the Ethernet Shield dynamically, and then the sketch sends the connection details to the serial port. You can view this information in the serial port monitor in the Arduino IDE.

```
#include <SPI.h>
#include <Ethernet.h>

byte mac[] = { 0x00, 0xC3, 0xA2, 0xE6, 0x3D, 0x57 };

void setup() {
  //D53 on an Arduino Mega must be an output.
  pinMode(53, OUTPUT);

  Serial.begin(9600);
  while (!Serial);

  Serial.print("Establishing network connection... ");

  if (Ethernet.begin(mac) == 0) {
    Serial.println("FAILED!");
  }
  else {
    Serial.println("OK!");

    Serial.print("IP Address: ");
    Serial.println(Ethernet.localIP());

    Serial.print("Default Gateway: ");
    Serial.println(Ethernet.gatewayIP());

    Serial.print("Subnet Mask: ");
    Serial.println(Ethernet.subnetMask());

    Serial.print("DNS Server: ");
    Serial.println(Ethernet.dnsServerIP());
  }
}

void loop() {
```

```
}
```

The method *begin()* in the Ethernet library's *Ethernet* class attempts to connect to the network using the details passed into it as arguments.

There are actually four forms of this method that you can use, depending on how much information you want to specify:

```
void begin(uint8_t *mac, IPAddress ip)
void begin(uint8_t *mac, IPAddress ip, IPAddress dns)
void begin(uint8_t *mac, IPAddress ip, IPAddress dns, IPAddress gateway)
void begin(uint8_t *mac, IPAddress ip, IPAddress dns, IPAddress gateway, IPAddress subnet)
```

At a bare minimum, you must call *begin()* and pass a MAC address as an array of bytes. If you declare the MAC address with the keyword const, you will need to cast it to a pointer of *uint8_t* values. For example:

```
Ethernet.begin((uint8_t*)mac);
```

If you pass an IP address then the Ethernet Shield will make a network connection using a static IP address. If you do not define an IP address then the shield will obtain one from the router using DHCP. For more information about static IP addresses, see Using a Static IP Address on page 66.

The remaining two, optional parameters are usually not needed except when working with complicated networks. If you have to specify the gateway address then you must use a static IP address. If you need to specify the subnet address then you must provide all three of the other arguments, and connect to the network using a static IP address.

begin() will return the value *1* if it connected successfully, and *0* if the connection failed.

If the sketch fails to establish a connection then there are a few things to try:

- Change the code to use a different MAC address.
- Check that the link indicator (as described above) is solid green or blinking. If there is no light then this indicates a problem with your wiring.
- Check your wiring carefully. Replace the Ethernet cable (if possible), and try a different connection method – such as directly to your router.

When the sketch successfully establishes a network connection, it calls four methods of the *Ethernet* class to retrieve the configuration settings that were given by the router.

Method	Description
dnsServerIP()	Returns the IP address of the primary DNS server used to lookup domain names to find their IP addresses.
gatewayIP()	Returns the IP address of the router device that provides the others with network access.
localIP()	Returns the IP address of the Ethernet Shield on the network.
subnetMask()	Returns the subnet mask used by the network to segregate devices into logical groups.

The DNS server, gateway, and subnet mask details sent to the serial port should match those used by other devices on your network.

To learn how to set the IP address and other network configuration parameters manually, and related topics such as port forwarding and dynamic DNS, see Arduino as a Web Server on page 65.

Testing the Connection

To check that the Arduino is able to respond to network traffic, you can ping it.

On Windows 8/7/Vista/XP:

1. Press the **Windows logo key + R**.
2. Type *cmd*, then press **Enter**.
3. Type *ping*, followed by a space, and then the IP Address displayed in the serial port monitor of the Arduino IDE.
4. Press **Enter**.

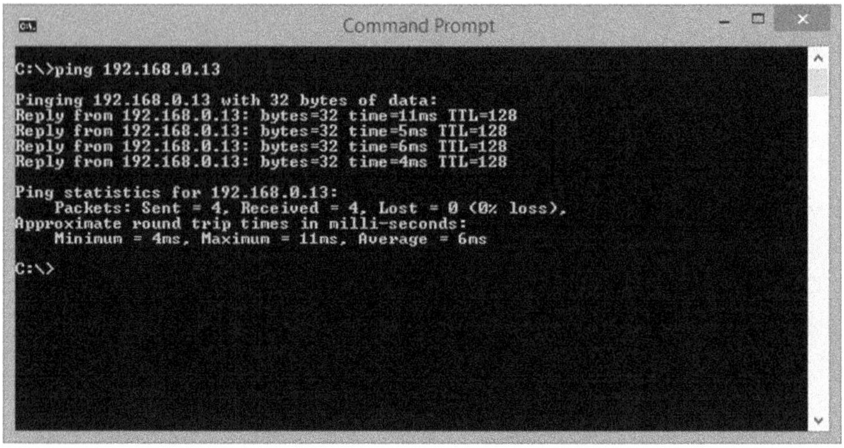

Figure 4. A successful ping on Windows

On Mac OS X:

1. On the dock, click **Finder**.

2. On the sidebar, click **Applications**.

3. Click **Utilities**, then double-click **Network Utility**.

4. On the **Ping** tab, in the box labelled "Enter the network address to ping", type the IP Address displayed in the serial port monitor of the Arduino IDE.

5. Click the **Ping** button.

Introducing Web Clients and Web Servers

On computer networks, such as the Internet, machines and devices can be defined in terms of the roles they play when exchanging information. There are two different roles: clients and servers.

Clients start connections with other machines in order to access the information that is contained on them. For example, a web browser is a client that connects to other machines to request web pages or files from them.

The machine that stores and "serves" the information that is requested is known as a server. Servers sit and wait until a client starts a conversation with them, and they are typically capable of talking to many thousands of clients at the same time. If the server is "offline", or it is in any way inaccessible, then the information that needs to be shared with clients is unavailable.

In most modern computer systems, a machine can act as both a server and a client at the same time. Its role in the exchange of information depends on the software it is running, and it can run both client software and server software if you need it to. But the Arduino's relatively low hardware specifications, and its lack of a multi-tasking operating system, largely prevent it from being both a server and a client at the same time.

When writing Arduino sketches that use the Ethernet Shield, you will often need to decide between programming a client and programming a server. The questions to ask are:

1. Do you need to send requests to other systems, for example Twitter, eBay, Facebook, and others to accomplish the task? If you do, you will create a client because those systems are servers.

2. Does the Arduino gather or hold information that multiple other machines need to access? If so, you will create a server.

3. Does the Arduino start connections, or should it sit and wait for others to connect to it? Create a client when you need to start the connections, and a server when you need to wait for incoming connections to be created by another device.

For information about creating clients using the Arduino Ethernet Shield, see Arduino as a Web Client on page 29.

For information about creating server projects using the Arduino Ethernet Shield, see Arduino as a Web Server on page 65.

Using SD Cards

The Arduino has quite a small amount of on-board storage and memory. On its own, it cannot store enough information to serve a large web-based interface or send many files to connected clients. And when acting as a client itself, many of the files an Arduino project needs to download are too big to be held in memory.

The Arduino Ethernet Shield comes with a built-in Secure Digital (SD) card socket that you can access using the SD library, *SD.h*. This library is supplied with the Arduino integrated development environment (IDE), and supports FAT16 and FAT32 file systems on standard SD cards and high-capacity SDHC cards.

SD.h is a wrapper that simplifies access to the SD card. It uses another library, *SDFat.h*, which is not covered in this book. *SDFat.h* is much more complicated, containing many methods and data structures for working with the SD card at a low-level, and it may be interesting to readers who are already experienced with SD cards.

In This Chapter

Formatting and Initializing SD Cards

The Ethernet Shield's built-in microSD socket supports FAT16 and FAT32-formatted cards through the SD library. Most SD and SDHC microSD cards will be formatted this way by the manufacturer. Extended-capacity SDXC cards (which usually come in capacities above 32GB) are not supported.

If you intend to use an SD card that has already been used in another device, which may have formatted it with a file system other than FAT16 or FAT32, you should format the card. There is no method or library supplied with the Arduino IDE to do this for you, and you will need to use a PC.

To format the SD card on Windows 8/7/Vista/XP:

1. Insert the SD card into a suitable card socket or USB card reader.

2. Press the **Windows logo key + R**. Type *explorer* and press **Enter**.

3. From the Windows File Explorer window, right-click the card device (usually labelled "SDHC" or "Removable Disk") and then click **Format**.

4. In the **File system** list, click **FAT**.[1]

5. In the **Allocation unit size** list, click **Default allocation size**.

6. Click **Start**.

Figure 5. Formatting SD cards on Windows

---

1. FAT32 should also work without any problems.

On Mac OS X:

1. Insert the SD card into a suitable card socket or USB card reader.

2. On the dock, click **Finder**.

3. On the sidebar, click **Applications**.

4. Click **Utilities**, and then double-click **Disk Utility**.

5. In the left panel, click the SD card (usually labelled "NO NAME" if the card was not formatted with a volume name).

6. On the **Erase** tab, from the **Volume Format** list, click **MS-DOS File System** or **MS-DOS(FAT)**.[1]

7. Click **Erase**.

Initializing SD Cards in Arduino Sketches

Before reading from an SD card, you need to include *SD.h* in the sketch. SD cards are serial peripheral interface (SPI) devices, and so you must also include the SPI library.

```
#include <SPI.h>
#include <SD.h>
```

You do not need to include the Ethernet library if you are only using the SD card socket and not actually connecting to a network.

The next step is to initialize the card using *SD.begin(4)*. This method accepts one argument and that is the pin number for the slave select function. Multiple SPI devices can share most of the SPI connections wires, but each device must have its own slave select wire. On the official Arduino Ethernet Shield, slave select for the SD card is digital pin 4.

If you are using an Ethernet Shield clone from a different manufacturer, the slave select function might be a different pin. You will need to locate this pin and change the code accordingly.

SD.begin() is a simplified version of the initialization process, and it not only initializes the SD card but also the FAT file system. If there is an error in either of those stages then the function returns false.

```
#include <SPI.h>
#include <SD.h>

void setup() {
  Serial.begin(9600);
```

---

1. The available options will depend on which version of Mac OS X you are running.

```
  while (!Serial);

  //D53 on an Arduino Mega must be an output.
  pinMode(53, OUTPUT);

  Serial.print("Initializing SD card... ");
  if (!SD.begin(4)) {
    Serial.println("FAILED!");
  }
  else {
    Serial.println("OK!");
  }
}
void loop() {
}
```

There are a number of reasons why card initialization can fail – including if there is not a card in the socket. If the SD card works on a PC then it generally safe to assume that there is something in the file system that the Arduino SD library cannot support. Try reformatting the SD card as described above.

> **Tip: Initializing the FAT file system can sometimes fail if there are no files on the card. Try reformatting the card and then transferring a file to it from your PC. Any file should be fine, provided that it is not 0 bytes long. Remember to eject the SD card from your PC before removing it.**

Reading from SD Cards

SD.h contains no method for directly listing the contents of the SD card, and the file system entries that contain this information are inaccessible unless you use *SDFat.h* to initialize the card. So to get information about the files and directories on the card, you must open each item one-by-one using the *SD.open()* method.

First, to open the card's top-level directory, use the call:

```
File fp = SD.open("/");
```

The *open()* method returns an instance of the *File* class. This contains methods for reading and writing files on the SD card, and other methods that can be used for reading information about those files.

It is very important that you close files and directories when the code does not need them anymore. To close a file or directory, call the method *close()* of the open file's instance:

```
fp.close();
```

The *File* class can also open files and directories without you specifying the item's name – using *openNextFile()* – by finding the next item based on the file and directory entries in the file system.

You can use this, and other methods from the *File* class, to write code that displays a file list similar to a command line *dir* or *ls* operation. The methods of the *File* class that will be used are:

Method	Description
name()	Returns the name of the current file.
openNextFile()	Opens the next file in a directory. Evaluates to *false* if the end of the directory has been reached.
isDirectory()	Returns *true* if the open file is a directory, and *false* if it is a data file.
size()	Returns the size of the file, in bytes.
close()	Closes a file.

More information and examples of how to list files can be found in Arduino as a Web Server on page 65.

```
#include <SPI.h>
#include <SD.h>

void setup() {
  Serial.begin(9600);
  while (!Serial);

  //D53 on an Arduino Mega must be an output.
  pinMode(53, OUTPUT);

  Serial.print("Initializing SD card... ");
  if (!SD.begin(4)) {
    Serial.println("FAILED!");
  }
  else {
    Serial.println("OK!");

    File root = SD.open("/");
    Serial.println(root.name());

    File lsf;
    while ((lsf = root.openNextFile())) {
      if (lsf.isDirectory()) {
        Serial.print("<DIR>       ");
      }
      else {
        char fsize[13];
        sprintf(fsize, "%10d  ", lsf.size());
        Serial.print(fsize);
      }
      Serial.println(lsf.name());
      lsf.close();
    }
    root.close();
```

```
  }
}
void loop() {
}
```

> **Tip: The library used to process the FAT file system only supports file names in the 8.3 format – eight characters for the name, a period, and then three characters for the file extension. It automatically converts file names that do not follow this format.**

Reading from a File

Use the method *SD.open()* to open a file when you know its name. Files that are in the top-level directory on the SD card can be opened by passing a string value containing the file name as the first (or only) argument. To open files in subdirectories, include the full path to the file in the string. For example: *folder/folder/file.ext*.

The instance of the *File* class returned by a call to *SD.open()* will evaluate to *false* if the file cannot be opened, or if it cannot be found. To check only whether a file exists, use the method *SD.exists()*, which returns *true* if the file is there or *false* if the file cannot be found.

```
Serial.print("Checking for readme file... ");
if (SD.exists("README.TXT")) {
  Serial.println("OK!");
}
else {
  Serial.println("NOT FOUND!");
}
```

Once you have opened a file with *SD.open()*, there are two methods of the *File* class that you can use to read bytes from the file: *read()* and *peek()*.

If you do not pass any arguments, a call to *read()* returns the next byte from the file, and advances your position in the file. The method *available()* returns the number of bytes left in the file that you have not yet read, and this can be used to detect the end of the file. Once your position reaches the end of the file, there are no bytes available.

To read the entire file from the SD card, you can loop until *available()* returns zero, reading and printing bytes one-by-one:

```
#include <SPI.h>
#include <SD.h>

void setup() {
  Serial.begin(9600);
  while (!Serial);

  //D53 on an Arduino Mega must be an output.
  pinMode(53, OUTPUT);
```

```
    Serial.print("Initializing SD card... ");
    if (!SD.begin(4)) {
      Serial.println("FAILED!");
    }
    else {
      Serial.println("OK!");

      Serial.print("Checking for readme file... ");
      if (SD.exists("README.TXT")) {
        Serial.println("OK!");
        Serial.println();

        File rm = SD.open("README.TXT");
        while (rm.available() > 0) {
          Serial.write(rm.read());
        }
        rm.close();

        Serial.println();
      }
      else {
        Serial.println("NOT FOUND!");
      }
    }
  }

void loop() {
}
```

Peek() can be useful occasionally, but it is not used by the examples in this book. It reads the next byte from the file in the same way as *read()*, but does not advance your position in the file.

An alternate form of *read()* is used to read multiple bytes into an area of memory – often called a buffer – including the memory occupied by an array or a struct, or dynamically-allocated with *malloc()*. When used in this way, *read()* accepts two arguments and returns an integer value indicating how many bytes were read from the file:

```
int read(void *buf, uint16_t nbyte);
```

buf is a pointer to an area of memory. The name of an array is also a pointer to an area of memory and can be used as this argument.

nbyte specifies the number of bytes that should be read from the file.

When working with arrays, keep the size of the data types in mind. An array of integer types defined as *int buf[100]* allocates memory that is 200 bytes long, as integer types on the Arduino are 16-bit values. The actual size in bytes of any type (including arrays and structs) can be calculated using the *sizeof()* operator.

After reading, the *read()* method moves your position in the file forward based on the number of bytes that it actually read from the file.

Working with Large Files

When working with large files, it is often not possible (or desirable) to load the entire file into the relatively small amount of memory available on the Arduino. It is preferable to read and process large files in chunks.

One way that you can do this is to use *read()* to fetch a limited number of bytes from a file and store them temporarily in an array or block of memory. After you have processed the first chunk, call *read()* again to load the next chunk into the same memory buffer, replacing the original data. The *File* class keeps track of your position in the file and ensures that you do not read the same data twice.

However, as an example, if you are writing a function that searches for a keyword in all of the files on the SD card then reading into a buffer in the way described so far may not work. If you have a file that contains ten ASCII characters, A–J, and that the size of the buffer is five characters, then this will divide the file processing into two chunks – ABCDE and FGHIJ. If you are looking for the sequence DEF in each buffer, you will not find it. DEF is not contained in the first chunk, nor is it contained in the second chunk.

One solution to this is to overlap reads from the card – moving backwards a few characters after reach call to *read()*. When searching for a sequence of three characters, the maximum amount that you need to move backwards is two bytes. This produces the chunks ABCDE, DEFGH, and GHIJ. DEF can now be found in the result of the second read operation.

The *File* class contains methods to help with moving backwards and forwards through the file, without actually reading values.

Method	Description
position()	Returns a 32-bit integer indicating the current position in the file.
seek()	Moves to a new position in the file.

seek() accepts one argument and that is a 32-bit integer ranging from zero (the start) to the size of the file (the end). This argument represents the absolute position to move to. To step backwards two bytes from the current position, use a call like:

```
fp.seek(fp.position() - 2);
```

To step forwards two bytes from the current position, you can use:

```
fp.seek(fp.position() + 2);
```

Using *seek()* frequently will significantly increase how long it takes for your sketch to process files. Using larger buffers will help, but you should always try to balance how fast you need the sketch to run against how much memory it uses.

Writing to SD Cards

To write to a file on the SD card, you open it using a different mode. But the other aspects of initializing the card are the same as described in Reading from SD Cards on page 21.

When opening files, the default action is to open them in read-only mode. An optional second argument to *SD.open()* sets the mode to one of the following values:

Mode	Description
FILE_READ	Opens the file in read-only mode. This is the default action.
FILE_WRITE	Opens a file for writing. If the file does not exist, it will be created.

When a file is opened with FILE_WRITE and it already exists, it is opened at the end so that any data that you write is appended to the existing content. If you want to overwrite a file then the easiest way is usually to delete the existing file before the call to *SD.open()*.

The instance of the *File* class that is returned by a successful call to *SD.open()* can be used to write to the file, using the method *write()*. Like *read()*, *write()* also has two forms – the first accepts a single byte argument, and this byte is written to the file in the current position. The second form accepts a pointer to the area of memory that contains the data to be written to the file, and an integer number specifying how many bytes are to be written. The name of an array is actually a pointer and so this can be used when calling *write()*.

There is no guarantee that the data will be written to the file immediately – it may only be saved when the file is closed. To ensure that the data is written, you can call the method *flush()*, which writes any information that is still held in the Ethernet Shield's buffers.

The following code sample creates a new file, and then writes the number 65 to it. When the file is opened with a text editor, such as Notepad, TextEdit, or Vi, this number appears as the ASCII character 'A'. The code then writes an array of numbers, which appears in a text editor as BCDEF.

Finally, this sample demonstrates how calls to *write()* can be used to store entire structures with one call.

```
#include <SPI.h>
#include <SD.h>

struct TS {
  byte G;
  byte H;
  byte I;
};

void setup() {
  Serial.begin(9600);
  while (!Serial);

  //D53 on an Arduino Mega must be an output.
  pinMode(53, OUTPUT);

  Serial.print("Initializing SD card... ");
  if (!SD.begin(4)) {
    Serial.println("FAILED!");
  }
  else {
    Serial.println("OK!");

    if (SD.exists("TEST.TXT"))
      SD.remove("TEST.TXT");

    Serial.print("Opening file for writing... ");
    File test = SD.open("TEST.TXT", FILE_WRITE);
    if (!test) {
      Serial.println("FAILED!");
      return;
    }
    Serial.println("OK!");

    // Write a byte
    test.write(65);

    // Declare an array and then write it to the file
    byte buf[5] = { 66, 67, 68, 69, 70};
    test.write(buf, 5);

    // Create a struct and then write it to the file
    TS myTest = {71, 72, 73};
    test.write((uint8_t*)&myTest, sizeof(TS));

    // Close the file
    test.close();

    Serial.println("Finished!");
  }
}

void loop() {
}
```

Deleting Files

Deleting a file is a form of writing, during which entries are written to the FAT file system indicating that a file has been deleted and that the areas of the SD card that it occupies can be overwritten.

To delete a file, use the method *SD.remove()* and pass the file name (with file path if needed) as a string argument:

```
SD.remove("DELETEME.TXT");
```

Making and Removing Directories

To create a directory, use the method *SD.mkdir()* and pass a string containing the name of the directory to create. The file path can be included in this string if the directory is to be created inside another directory.

```
SD.mkdir("DIR_NAME");
```

If any intermediate directories do not exist, this method will create them too. For example, passing the string *TEMP1/TEMP2* to *SD.mkdir()* will create two new directories – the first being TEMP1, and the second being TEMP2, which is created inside TEMP1.

To delete a directory, which also deletes any files and subdirectories that it contains, use the method *SD.rmdir()*:

```
SD.rmdir("DIR_NAME");
```

Arduino as a Web Client

Using the *EthernetClient* class in the Arduino Ethernet library, the sketches you run on an Arduino can connect to servers on the Internet and your local network, and download information from them. These types of projects are primarily concerned with initiating connections, sending requests for files or information, and processing responses from the server.

This chapter guides you through the processes and techniques involved in downloading information from the Internet to your Arduino. You do not have to complete all three projects but, as each project is designed to build-on the information presented in the preceding one, it is recommended that you at least read through them.

The code samples in the following sections are examples of how to communicate over hypertext transfer protocol (HTTP). HTTP is outlined in full in Hypertext Transfer Protocol – HTTP/1.0 on page 144, but you only need a small amount of background information to make the code work. You will learn about the key parts of HTTP as you progress through the projects.

In This Chapter

Project 1 – Setting up a Basic Web Client

Chapter 1 shows how to connect the Arduino Ethernet Shield to an Arduino and make a network connection from a sketch. The mechanics of how a line of communication is maintained between two machines over a network, and how messages are delivered, are covered by the transmission control protocol and Internet protocol (TCP/IP).

Ports allow different types of messages to be received by different pieces of software running on the same server. For example, server software that accepts connections over File Transfer Protocol (FTP) will usually run on port 21. Web server software usually accepts connections over HTTP on port 80. Depending on the port that a connection is made on, the server can route the messages through to the correct piece of software.

The *EthernetClient* class from the Arduino Ethernet library contains methods to help you establish connections with servers, and it deals with most of the complexity on your behalf.

To make requests and receive information from a web server, your Arduino needs to open up a line of communication with the server and then send it a request message. This request contains, at a minimum, the details of the file that you want to download. The server then sends back a response message, which includes the requested file and also information about that file.

The content of the requests and responses must follow an agreed structure so that each machine can understand the messages from the other.

In this project you will write an Arduino sketch that:

1. Connects to the network in the same way as described in Chapter 1.
2. Creates an HTTP request message that tells a web server which file you'd like to download.
3. Gets the response message from the web server.
4. Saves the downloaded file to an SD card.
5. Repeats this process every 60 mins to keep the file up to date.

Starting the Sketch

To begin, start a new Arduino sketch and paste in the following the code, or type it carefully:

```
#include <SPI.h>
#include <Ethernet.h>

byte mac[] = { 0x00, 0xC3, 0xA2, 0xE6, 0x3D, 0x57 };
byte LED = 2;

void setup() {
  //D53 on an Arduino Mega must be an output.
  pinMode(53, OUTPUT);

  pinMode(LED, OUTPUT);

  Serial.begin(9600);
  while (!Serial);

  Serial.print("Establishing network connection... ");

  if (Ethernet.begin(mac) == 0) {
    Serial.println("FAILED!");

    // signal that there was a network problem, and wait for reset.
    while (true) {
      digitalWrite(LED, HIGH);
      delay(500);
      digitalWrite(LED, LOW);
      delay(500);
    }
  }
  else {
    Serial.println("OK!");
    digitalWrite(LED, LOW);
```

```
  }
}
void loop() {
}
```

If you would like more information about initializing the network and using the test sketch, see Establishing a Network Connection on page 12. The code above is essentially the same as used in Chapter 1 – the only notable difference is that a light-emitting diode (LED) is connected to the Arduino on digital pin 2 (through a 220Ω resistor) and this is used as a visual alarm if there is a problem making a connection to the network.

If the sketch cannot make a connection to the network, the *while(true)* loop flashes the LED until the Arduino is reset. This prevents the sketch from progressing to the *loop()* function.

> **Tip: Remember that digital pin 13 is used by the Ethernet shield and so the Arduino's built-in LED on that pin cannot be used.**

Connecting to a Web Server

At the top of the sketch, underneath the MAC address and LED pin declaration, declare an instance of the Ethernet library's *EthernetClient* class:

```
EthernetClient myClient;
```

And then define two character arrays that specify the domain name of the web server you wish to connect to, and the file that you want to download:

```
char wServer[] = "www.arduino.cc";
char wFile[] = "/";
```

These two strings form part of a web address, a universal resource locator (URL). The full URL when accessed from a regular web browser on your PC is *http://www.arduino.cc/*

When used this way in an Arduino sketch:

- • You do not need to specify that you're using HTTP in the server name.
- • Requesting the file "/" is the same as asking for the website's default page.

Enter the following code as the sketch's *loop()* function:

```
void loop() {
  Serial.print("Connecting to ");
  Serial.print(wServer);
  Serial.print("... ");
```

```
  if (myClient.connect(wServer, 80) == 1) {
    Serial.println("OK!");
    myClient.stop();
  } else {
    Serial.print("FAILED!");
  }

  delay(600000);
}
```

The call to *delay()* is to ensure that this project waits for around 10 mins before it makes another request to the server.

Two methods of the *EthernetClient* class are introduced in this code: *connect()* and *stop()*.

The call to *connect()* opens a connection to a machine on the Internet (or local area network). The example code passes two arguments into the method: the first of these is the machine's domain name as an array of characters. The second argument is the port number (usually 80 for HTTP). If the call was successful then *connect()* returns the value *1*; if it returns anything else then there has been an error.

There are many reasons that calls to *connect()* can fail. If this happens:

1. Check that you can connect to your router using the sketch shown in Establishing a Network Connection on page 12.
2. Ensure you have typed the server name correctly in the sketch.
3. See if the webpage can be accessed using a web browser on your PC.
4. Reset the sketch or wait until the *delay()* expires and the code attempts another connection. Some problems are only temporary.

Once you have finished working with a connection, you should always close it as the Ethernet Shield can only support four simultaneous connections. To close the connection, call the method *stop()* from the active instance of the *EthernetClient* class.

In this project, you set the server's domain name using a character array. When the *connect()* method is called, it translates this domain name into an IP address for you. Earlier versions of the Arduino Ethernet library did not do this.

If you want to connect by specifying an IP address instead, you can declare an instance of the *IPAddress* class and pass this into *connect()* instead of the character array:

```
IPAddress ip(174,129,243,245);
```

Sending an HTTP Request

Now that you have made a connection to a web server, you will extend the code above to send an HTTP request. HTTP requests are nothing more than strings of characters that are sent to the server to tell it what content you would like.

There are two types of HTTP request: GET and POST. At this stage, the difference between them is not important and you will use GET[1]. Here is an example of a GET request:

```
GET / HTTP/1.0[crlf]
Host: www.arduino.cc[crlf]
Connection: close[crlf]
[crlf]
```

The line breaks above are intentional and are encoded in HTTP request and response messages as two-byte sequences – ASCII character 13 (carriage return) followed by ASCII character 10 (line feed).

The first line is called the *request line*. After the keyword GET, there is the file name and file path of the information that you are requesting. In this case, the variable *wFile* is inserted and this currently contains the value "/".

On the same line, the characters "HTTP/1.0" show the version of the HTTP protocol[2] that the request conforms to, and the version that the server should use when responding.

Host and *Connection* are two header fields that you can choose to send. These are like arguments passed into a function – they specify additional information that the web server can use to fulfil the request.

Host simply restates the domain name of the server that you are contacting – some servers host many websites and need you to put the website that you are wanting to talk to in the HTTP request. *Connection* specifies the type of HTTP connection to be used. In this case, *close* tells the web server to terminate the connection once it has responded to the request. Neither *Host* nor *Connection* are actually part of the HTTP/1.0 protocol, they are from HTTP/1.1. However, communication with many web servers relies on them being included. You can find descriptions of the acceptable headers in HTTP/1.0 in the appendix, section 5.2 Request Header Fields on page 161.

---

1. The HTTP POST method is used in Project 3 – Building a Twitter Alarm on page 51.
2. The current version of HTTP is version 1.1. However, HTTP/1.1 includes several features that make it slightly more difficult for small web clients to deal with.

Send information to the server using the *print()*, *println()*, or *write()* methods of the *EthernetClient* class. These methods work in the same way as their counterparts used to send data to Arduino's serial port. To generate and send the HTTP request shown earlier in this section, insert the following code between the lines *Serial.println("OK!")* and *myClient.stop()* in the sketch's *loop()* function:

```
myClient.print("GET ");
myClient.print(wFile);
myClient.println(" HTTP/1.0");
myClient.print("Host: ");
myClient.println(wServer);
myClient.println("Connection: close");
myClient.println();
```

The empty line sent to the server at the end of this code tells the web server that you have finished sending the HTTP request and that it should now respond.

There is no observable difference if you run the sketch at this point. The current code in the sketch's *loop()* closes the connection without waiting for the web server's response.

Getting the Server's Response

To continue, you are going to add code to receive the web server's response and save the information to a file on the SD card.

Immediately after the line *myClient.println()* in the sketch's *loop()* function, add the following code:

```
if (SD.begin(4)) {
  if (SD.exists("DOWNLOAD.TXT"))
    SD.remove("DOWNLOAD.TXT");

  Serial.print("Saving response... ");
  digitalWrite(LED, HIGH);
  File dd = SD.open("DOWNLOAD.TXT", FILE_WRITE);

  while (myClient.connected()) {
    if (myClient.available() > 0) {
      dd.write(myClient.read());
    }
  }

  dd.close();
  delay(500);
  Serial.println("OK");
  digitalWrite(LED, LOW);
}
else {
  Serial.println("No SD card detected!");
}
```

If you are not familiar with working with the Arduino Ethernet Shield's built-in SD card socket and the *SD.h* library, you may wish to read Using SD Cards on page 18.

Three new methods of the *EthernetClient* class are included in the code above:

Method	Description
connected()	Returns *true* if the connection to the server is still active, or *false* if it is no longer usable.
available()	Returns the number of bytes that are currently in the Ethernet Shield's buffer, waiting to be read.
read()	With no arguments, *read()* fetches a single byte from the server's response and returns it.

After creating a new file to hold the data, the code enters a while loop that waits until the server drops the HTTP connection. Since there is no guarantee that all of the data will be sent immediately, this loop waits for available data and then writes it to the SD card when it does arrive.

Using HTTP/1.0, the server should end the connection once it is finished sending its response. Your HTTP request includes the field *"Connection: close"* to help ensure that this happens when communicating with HTTP/1.1 servers. However, one of the most common causes of problems is web servers not closing connections. Downloading files from the Internet through the Ethernet Shield can be slow, but if the LED remains lit for a very long period of time then it may mean that the server has not closed the connection.

In the next project, Project 2 – Scraping Webpages to Retrieve Information on page 39, you will see how to implement timeouts so that the sketch can recover if the server stops sending information but does not close the connection. For now, if you encounter this problem, try downloading a different file from a different web server.

Once the Arduino has downloaded the file and the LED is turned off, you can turn off the Arduino and remove the SD card. Insert the SD card into your PC and open up the file *DOWNLOAD.TXT* in a text editor such as Notepad, TextEdit, or Vi. The file contains HTML code for the webpage you requested, but with several lines of information before that.

Figure 6. An HTTP response from a web server

Beginning with "HTTP/1.1" and ending with the blank line, this is the server's HTTP response header and it is not a part of the webpage. The first line of this response is the most important, and it is called the *status line*.

Even though your code requests that the server use HTTP/1.0, this particular web server responds with HTTP/1.1. However, it does not use any of the HTTP/1.1 features that could make it difficult for you to process the response.

After the protocol version, the server sends an HTTP status code (200) and a *reason phrase* that describes the code. A full list of HTTP status codes is shown in Appendix A, section 9. Status Code Definitions on page 166. However, for the purpose of checking that the HTTP request completes successfully, there are only a few status codes that you need to be aware of:

HTTP Status Code	Description
200	OK. The request completed successfully.
400–499	Indicates that the request failed because of a problem caused by the client. For example, requesting a file that is not on the server (404) or making an invalid request.
500–599	Indicates that the request failed because of a problem on the server.

Regardless of the status code, most web servers return an HTML page after the HTTP response header. This page may not be the file you requested and, instead, may be a page that describes the error that has occurred.

In this project you will only modify the sketch so that it ignores the HTTP header and does not write it to the file. This means that the file saved to the SD card could be an error page.

> **Tip: You can use a web-based tool such as Rex Swain's HTTP Viewer (http://www.rexswain.com/httpview.html) to verify how a web server responds to an HTTP/1.0 request.**

Before the *while(myClient.connected())* loop, add the following lines:

```
char lc;
while (myClient.connected()) {
  if (myClient.available()) {
    char nc = myClient.read();
    if ((lc == 10) && (nc == 13)) {
      while (myClient.available() == 0);
      myClient.read();
      break;
    }
    else
      lc = nc;
  }
}
```

This is not pretty code, but it works. The loop skips characters from the server's response until it finds a line feed (10) followed by a carriage return (13) – it matches the middle part of the sequence that is used to mark the end of the HTTP response header. The code then skips the next character (the final line feed) and terminates the loop.

The sketch can now write the remaining characters from the server's response to the SD card. In this project, the LED is lit while it writes the file so that you know not to remove the card in the middle of the process.

Resetting the Sketch

Unfortunately, this project highlights one of the major limitations in the Arduino's SD library, *SD.h*: it cannot detect when the SD card is removed. This means that attempts to work with files on the SD card may appear to succeed, even after the card is removed.

You cannot even make repeated calls to *SD.begin()* to detect the presence of a valid card after it initially detects an SD card.

This is not a problem if you turn off the Arduino before removing or inserting the SD card. However, this is a more usable project if it can support the user removing the SD card without turning off the power.

Restarting the sketch on the Arduino re-initializes the data used by SD.h, so that the presence of the card can be properly detected again. This leads to the hack at the end of the sketch's *loop()* function. After the delay, the following piece of inline assembly language resets the sketch:

```
asm volatile (" jmp 0");
```

Source Code

The full source code for the file-downloading sketch is shown below.

This sketch has been changed to save something more useful than the Arduino homepage to the SD card. Since downloading this file too often is not especially useful, the delay in this sketch has been extended to wait for approximately 1 hour.

```
#include <SPI.h>
#include <Ethernet.h>
#include <SD.h>

byte mac[] = { 0x00, 0xC3, 0xA2, 0xE6, 0x3D, 0x57 };
byte LED = 2;
char wServer[] = "media.wizards.com";
char wFile[] = "/images/magic/tcg/resources/rules/MagicCompRules_20140601.pdf";

EthernetClient myClient;

void setup() {
  //D53 on an Arduino Mega must be an output.
  pinMode(53, OUTPUT);

  pinMode(LED, OUTPUT);

  Serial.begin(9600);
  while (!Serial);

  Serial.print("Establishing network connection... ");
  if (Ethernet.begin(mac) == 0) {
    Serial.println("FAILED!");
    while (true) {
      digitalWrite(LED, HIGH);
      delay(500);
      digitalWrite(LED, LOW);
      delay(500);
    }
  }
  else
    Serial.println("OK!");
}

void loop() {

  Serial.print("Connecting to ");
  Serial.print(wServer);
  Serial.print("... ");
```

```
  if (myClient.connect(wServer, 80) == 1) {
    Serial.println("OK!");

    myClient.print("GET ");
    myClient.print(wFile);
    myClient.println(" HTTP/1.0");
    myClient.print("Host: ");
    myClient.println(wServer);
    myClient.println("Connection: close");
    myClient.println();

    if (SD.begin(4)) {
      if (SD.exists("MAGIC.PDF"))
        SD.remove("MAGIC.PDF");

      Serial.print("Saving response... ");
      digitalWrite(LED, HIGH);
      File dd = SD.open("MAGIC.PDF", FILE_WRITE);

      char lc;
      while (myClient.connected()) {
        if (myClient.available()) {
          char nc = myClient.read();
          if ((lc == 10) && (nc == 13)) {
            while (myClient.available() == 0);
            myClient.read();
            break;
          }
          else
            lc = nc;
        }
      }

      while (myClient.connected()) {
        if (myClient.available() > 0) {
          dd.write(myClient.read());
        }
      }

      dd.close();
      delay(500);
      Serial.println("OK");
      digitalWrite(LED, LOW);
    }
    else
      Serial.println("No SD card detected!");

    myClient.stop();
  }
  else
    Serial.println("FAILED!");

  delay(3600000);
  asm volatile(" jmp 0");
}
```

Project 2 – Scraping Webpages to Retrieve Information

Web scraping is a programming technique for finding information on websites that do not have an application programming interface (API). It involves reading through the hypertext markup language (HTML) code that web browsers use when displaying webpages. Scraping finds information by looking at elements in the code that usually do not change when the page is updated, and using these elements to help find nearby information that is more likely to change.

In this project, you will write an Arduino sketch that reads a price from the online retailer GameStop.com. The sketch will read the pre-owned price of the item every 10 mins and, when the price is reduced, it will turn on a light-emitting diode (LED) to notify you that the item is now cheaper.

Figure 7 is a screenshot of the item that this sketch analyzes. You will be looking for the "BUY PRE-OWNED" price ($27.99) in the HTML.

Figure 7. Naruto Shippuden on Xbox 360 at GameStop.com

Starting the Sketch

To begin, start a new Arduino sketch and paste in the code below, or type it carefully. This is your starting point, and is based on the file-downloading sketch covered in Project 1 – Setting up a Basic Web Client on page 29. In its current state, the sketch connects to the network and sends an HTTP request to the server to fetch the webpage.

To keep the code a little tidier, this example places the code that fetches the HTML in a new function, and calls this function from the sketch's *loop()* function. It writes the HTML response from the server to the serial port, which you can view by using the serial port monitor in the Arduino integrated development environment (IDE).

The code to ignore the HTTP response header is now also contained in its own function – *skipHeader()*.

```
#include <SPI.h>
#include <Ethernet.h>

const byte mac[] = { 0x00, 0xBC, 0xA2, 0xE6, 0x3D, 0x57 };
const byte LED = 2;
const char wServer[] = "www.gamestop.com";
const char wFile[] = "/xbox-360/games/naruto-shippuden-ultimate-ninja-storm-3-full-burst/110541";

EthernetClient myClient;

void setup() {
  //D53 on an Arduino Mega must be an output.
  pinMode(53, OUTPUT);

  pinMode(LED, OUTPUT);

  Serial.begin(9600);
  while (!Serial);

  Serial.print("Establishing network connection... ");

  if (Ethernet.begin((uint8_t*)mac) == 0) {
    Serial.println("FAILED!");
    while (true) {
      digitalWrite(LED, HIGH);
      delay(500);
      digitalWrite(LED, LOW);
      delay(500);
    }
  }
  else {
    Serial.println("OK!");
  }
}

void loop() {
  if (getHTML())
    delay(600000);
  else
    delay(60000);
}

void skipHeader() {
  char lc;
  while (myClient.connected()) {
    if (myClient.available()) {
      char nc = myClient.read();
      if ((lc == 10) && (nc == 13)) {
        while (myClient.available() == 0);
        myClient.read();
        break;
      }
      else
        lc = nc;
    }
  }
}

boolean getHTML() {
```

```
Serial.print("Connecting to ");
Serial.print(wServer);
Serial.print("... ");

if (myClient.connect(wServer, 80) != 1) {
  Serial.println("FAILED!");
  return false;
}

Serial.println("OK!");

myClient.print("GET ");
myClient.print(wFile);
myClient.println(" HTTP/1.0");
myClient.print("Host: ");
myClient.println(wServer);
myClient.println("User-Agent: Mozilla/5.0 (Windows NT 6.3; WOW64; Trident/7.0; Touch: MAARJS; rv:11.0)
like Gecko");
myClient.println("Connection: close");
myClient.println("Cookie: user_country=USA");
myClient.println();

skipHeader();
while (myClient.connected()) {
    if (myClient.available() > 0) {
      Serial.write(myClient.read());
    }
}

myClient.stop();
return true;
}
```

Sending Additional HTTP Header Fields

Web browsers implement a lot of features so that they are able to display all websites and communicate with all web servers. When using the Ethernet Shield on an Arduino, you usually implement only the features that you need to make the project run.

Sometimes there can be a little trial and error involved in scraping public websites. In the case of GameStop.com, their server expects you to specify a user agent (a string that describes the client that is requesting the webpage), and accept a cookie so that your location is included with each request. You send this extra information to the server by using additional fields in the HTTP header of your GET request.

The starting point for this project specifies two additional fields so that it can work with GameStop.com.

It adds a fake[1] user agent by sending the string:
User-Agent: Mozilla/5.0 (Windows NT 6.3; WOW64; Trident/7.0; Touch; MAARJS; rv:11.0) like Gecko

---

1. This user agent string is not really fake – but it falsely identifies the Arduino as something else. For more information about user agents, see 10.15 User-Agent on page 178.

It rarely matters exactly what you send for this field, and borrowing one from a web browser is fine. A list of the user agents used by common web browsers is at *www.useragentstring.com*.

The HTTP request also sends a fake cookie, by adding the *Cookie* field to send a cookie called *user_country* with the value *USA*.

Reacting to HTTP Status Codes

In Sending an HTTP Request on page 33, you can see how the server responds to web requests, and the types of status codes it may return. In Project 1, and in the starting code for this project, the sketch skips the HTTP header and assumes that the request completed successfully. However, this may not always be the case.

Regardless of what the status of the response is, the three-digit code will begin at the 10th character returned by the server. So to retrieve the status code, the Arduino only needs to read the first 12 characters of the server's response.

You should write this part of the program in the *getHTML()* function before the call to *skipHeader()*.

You can use a second form of the *EthernetClient* class method *read()* to read multiple characters into a temporary buffer.

```
int read(uint8_t *buf, size_t size);
```

buf is a pointer to an area of memory. The name of an array is also a pointer to an area of memory and can be used as this argument.

size specifies the number of bytes that should be read.

To read past the HTTP protocol version and obtain the status code. First wait until the server sends, at least, the first 12 bytes of the response:

```
while (myClient.available() < 12);
```

Then read the first 12 bytes into a temporary array:

```
char buf[12];
myClient.read((uint8_t*)buf, 12);
```

And finally, extract the three-digit status code and convert it to an integer. The entries *buf[9]*, *buf[10]*, and *buf[11]* contain the three digits of the status code.

There are various ways of converting these three characters into a number. An example of such a function is shown here:

```
int getStatusCode(char sc1, char sc2, char sc3) {
  String tmp;
  tmp.concat(sc1);
  tmp.concat(sc2);
  tmp.concat(sc3);
  return tmp.toInt();
}
```

Add the *getStatusCode()* function to the sketch and then, after the statement *myClient.read((uint8_t*)buf, 12);* add the line:

```
int sc = getStatusCode(buf[9], buf[10], buf[11]);
```

At this point, the sketch should check that the HTTP response status code is 200.

Status codes in the range 300 through 308 are usually some kind of redirection. A web browser should, ideally, read through the remaining the fields in the HTTP header to find the new destination.

However, in this project, you only perform basic checks to see if the request completed successfully. If the status code is 300–499 then the sketch enters a while loop that flashes the LED and takes no further action. These error codes include files not being available at the specified location and errors that are caused by an invalid request to the server. It is unlikely that these problems will be fixed without making changes to the sketch and re-uploading it to the Arduino.

If the status code is 500–599 then the error might be temporary, and so the sketch is programmed to try fetching the webpage again a little later.

Only if the status code is 200 does the *getHTML()* function continue to read the data and process it. Since this project does not make any further use of the HTTP response header, make a call to *skipHeader()* to move past the header and up to the start of the HTML content.

At this stage, the *getHTML()* function should look like this:

```
boolean getHTML() {
  Serial.print("Connecting to ");
  Serial.print(wServer);
  Serial.print("... ");

  if (myClient.connect(wServer, 80) != 1) {
    Serial.println("FAILED!");
    return false;
  }

  Serial.println("OK!");
```

```
myClient.print("GET ");
myClient.print(wFile);
myClient.println(" HTTP/1.0");
myClient.print("Host: ");
myClient.println(wServer);
myClient.println("User-Agent: Mozilla/5.0 (Windows NT 6.3; WOW64; Trident/7.0; Touch: MAARJS; rv:11.0)
like Gecko");
myClient.println("Connection: close");
myClient.println("Cookie: user_country=USA");
myClient.println();

while (myClient.available() < 12);
char buf[12];
myClient.read((uint8_t*)buf, 12);
int sc = getStatusCode(buf[9], buf[10], buf[11]);

if ( (sc >= 300) && (sc <= 499)) {
  myClient.stop();
  while (true) {
    digitalWrite(LED, LOW);
    delay(500);
    digitalWrite(LED, HIGH);
    delay(500);
  }
}
else if ( (sc >= 500) && (sc <= 599)) {
  myClient.stop();
  return false;
}

// Will only reach here if status codes 200-226
// are received.
if (sc == 200) {
  skipHeader();
  Serial.println("OK! Looking for price...");
}

myClient.stop();
return true;
}
```

You can run the sketch. If all is well, the message "OK! Looking for price…" is sent to the serial port monitor.

Retrieving the Price

To find the price in the HTML code, you first need to identify a sequence of characters that indicate where the price begins. To do this, you need to examine the HTML code of the webpage.

Ideally, look for a sequence of characters that occurs only once in the document and is immediately followed by the price. This is different for every website, but it is usually the same for all similar webpages on the same site.

GameStop.com makes it straightforward: there are two locations for you to work with. The screenshot earlier shows the pre-owned price as it is displayed to the web browser. However, in addition to the code that makes up that part of the display, the site places the price in a short JavaScript code block.

```
<!-- landing_TrueTag -->
<script language="JavaScript">
var CI_SKU = '110541';
var CI_Category = 'Action';
var CI_Platform = 'Xbox 360';
var CI_Rating = 'T';
var CI_RegPrice = '29.99';
var CI_SalePrice = '';
var CI_PreOwnedPrice = '27.99';
</script>
<script src="http://cts.channelintelligence.com/11163_landing.js"></script>
<!-- end landing_TrueTag -->
```

The Arduino sketch can find the pre-owned price by waiting until it encounters the sequence of characters *var CI_PreOwnedPrice = '* and then reading until it finds the next apostrophe. In this sketch, the marker sequence is defined in the variable *mrkPrice*. The characters between the marker sequence and the next apostrophe are the price.

The process for extracting this value uses two *String* objects as temporary buffers. There are faster ways, but using *String* objects has the advantage of being slightly easier to understand for beginners.

Figure 8 shows the process that is implemented in the Arduino C code.

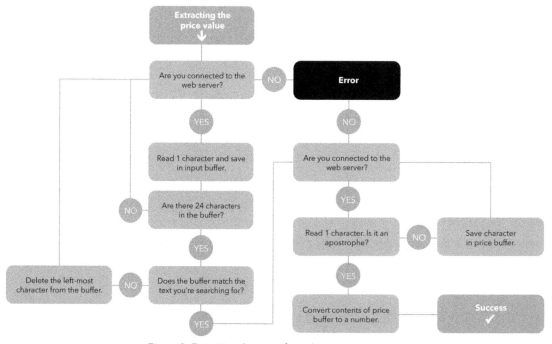

Figure 8. Extracting the price from the response stream

The equivalent Arduino C code for this process is constructed using two while loops:

```
String buffer = "";
String pBuf = "";
int bPtr = 0;

while (myClient.connected()) {
  if (myClient.available() > 0) {
    buffer.concat((char)myClient.read());
    bPtr++;
    if (bPtr == 24) {
      if (buffer == mrkPrice) {
        while (myClient.connected()) {
          if (myClient.available() > 0) {
              char tmp = myClient.read();
              if (tmp != '\'')
                pBuf.concat(tmp);
              else
                break;
          }
        }
        break;
      }
      buffer = buffer.substring(1);
      bPtr = 23;
    }
  }
}

float newPrice = stringToFloat(pBuf);
Serial.print("Price: $");
Serial.println(newPrice);
```

You can find the source code for the function *stringToFloat()* in the listing at the end of this project. If you are working with version 1.5 of the Arduino IDE (this version is in beta testing at the time of writing) then the *String* class now includes the method *toFloat()* that you can use instead.

Handling Timeouts

The GameStop.com server tends to work as expected – when responding to HTTP/1.0 clients, it closes the connection after it has sent its response. In addition, this sketch will also work with servers that mistakenly keep the connection open because it closes the connection to the server itself when it receives an unwelcome status code or when the price has been extracted successfully.

However, if the server returns the webpage successfully but the sketch cannot find the marker used to extract the price, then it is possible for the sketch to become stuck in a loop. If the server does not close the connection then the sketch might continue to wait for data which will never arrive.

There are several ways of building a timeout feature that stops the loop if no data is received for a long period of time. One way is to initialize an unsigned long variable and decrement it every time *available()* returns zero. When this timeout counter reaches zero, the sketch should break out of the loop.

Add a global unsigned long to the sketch:

```
unsigned long timeout;
```

In the function *getHTML()*, after declaring the two string buffers and the integer *bPtr*, initialize the timeout counter to a suitably high number. For example:

```
timeout = 60000;
```

Add the following code at the end of the main while loop in *getHTML()* so that it extends the statement *if(myClient.available() > 0) {}* with an *else* clause:

```
else {
  timeout--;
  if (timeout == 0)
    break;
}
```

Finally, after the same *if(myClient.available() > 0) {* statement, add this assignment to reset the timeout when data is received from the server:

```
timeout = 60000;
```

Completing the Project

Now that the sketch can extract the price and convert it to a float, it can compare the price with the initial value to see if the item is cheaper than it used to be. If the new value is less than the initial value, the LED is turned on.

Add a new global float variable to the sketch:

```
float itmPrice = 0.0;
```

And then add the following code after the line S*erial.println(newPrice)*:

```
if (itmPrice == 0)
    itmPrice = newPrice;
else if (itmPrice > newPrice)
    digitalWrite(LED, HIGH);
else
    digitalWrite(LED, LOW);
```

When the sketch first runs, *itmPrice* is initialized as zero. The price on the website will become the new benchmark. Since this also happens if the Arduino loses power, you could expand this project to save the benchmark price to the SD card and compare newly-retrieved prices against that instead.

Source Code

The complete source code for the price-monitoring sketch is shown below. When you have verified that everything is functioning correctly, you can safely remove all of the messages sent to the serial port and disconnect the Arduino from your PC. If powered and cabled to the network, it should monitor the webpage for as long as the Arduino is able to make a connection to the web server.

> **Tip: Arduinos can usually run for months without any problems. Do not be worried about leaving the device running.**

```
#include <SPI.h>
#include <Ethernet.h>

const byte mac[] = { 0x00, 0xB7, 0xA2, 0xE6, 0x3D, 0x57 };
const byte LED = 2;
const char wServer[] = "www.gamestop.com";
const char wFile[] = "/xbox-360/games/naruto-shippuden-ultimate-ninja-storm-3-full-burst/110541";
const String mrkPrice = "var CI_PreOwnedPrice = '";

float itmPrice = 0.0;
EthernetClient myClient;
unsigned long timeout;

void setup() {
  //D53 on an Arduino Mega must be an output.
  pinMode(53, OUTPUT);

  pinMode(LED, OUTPUT);

  Serial.begin(9600);
  while (!Serial);

  Serial.print("Establishing network connection... ");

  if (Ethernet.begin((uint8_t*)mac) == 0) {
    Serial.println("FAILED!");
    while (true) {
      digitalWrite(LED, HIGH);
      delay(500);
      digitalWrite(LED, LOW);
      delay(500);
    }
  }
  else {
    Serial.println("OK!");
  }
}

void loop() {
  if (getHTML())
    delay(600000);
  else
    delay(60000);
}
```

```
void skipHeader() {
  char lc;
  while (myClient.connected()) {
    if (myClient.available()) {
      char nc = myClient.read();
      if ((lc == 10) && (nc == 13)) {
        while (myClient.available() == 0);
        myClient.read();
        break;
      }
      else
        lc = nc;
    }
  }
}

boolean getHTML() {
  Serial.print("Connecting to ");
  Serial.print(wServer);
  Serial.print("... ");

  if (myClient.connect(wServer, 80) != 1) {
    Serial.println("FAILED!");
    return false;
  }

  Serial.println("OK!");

  myClient.print("GET ");
  myClient.print(wFile);
  myClient.println(" HTTP/1.0");
  myClient.print("Host: ");
  myClient.println(wServer);
  myClient.println("User-Agent: Mozilla/5.0 (Windows NT 6.3; WOW64; Trident/7.0; Touch: MAARJS; rv:11.0)
like Gecko");
  myClient.println("Connection: close");
  myClient.println("Cookie: user_country=USA");
  myClient.println();

  while (myClient.available() < 12);
  char buf[12];
  myClient.read((uint8_t*)buf, 12);
  int sc = getStatusCode(buf[9], buf[10], buf[11]);

  if ( (sc >= 300) && (sc <= 499)) {
    myClient.stop();
    while (true) {
        digitalWrite(LED, HIGH);
        delay(500);
        digitalWrite(LED, LOW);
        delay(500);
    }
  }
  else if ( (sc >= 500) && (sc <= 599)) {
    myClient.stop();
    return false;
  }

  // Will only reach here if status codes 200-226
  // are received.
  if (sc == 200) {
    skipHeader();

    String buffer = "";
    String pBuf = "";
    int bPtr = 0;

    while (myClient.connected()) {
```

```
    if (myClient.available() > 0) {
      timeout = 60000;
      buffer.concat((char)myClient.read());
      bPtr++;
      if (bPtr == 24) {
        if (buffer == mrkPrice) {
          while (myClient.connected()) {
            if (myClient.available() > 0) {
              char tmp = myClient.read();
              if (tmp != '\'')
                pBuf.concat(tmp);
              else
                break;
            }
          }
          break;
        }
        buffer = buffer.substring(1);
        bPtr = 23;
      }
    }
    else {
      timeout--;
      if (timeout == 0)
        break;
    }
  }
  float newPrice = stringToFloat(pBuf);
  Serial.print("Price: $");
  Serial.println(newPrice);

  if (itmPrice == 0)
    itmPrice = newPrice;
  else if (itmPrice > newPrice)
    digitalWrite(LED, HIGH);
  else
    digitalWrite(LED, LOW);
}

  myClient.stop();
  return true;
}

int getStatusCode(char sc1, char sc2, char sc3) {
  String tmp;
  tmp.concat(sc1);
  tmp.concat(sc2);
  tmp.concat(sc3);
  return tmp.toInt();
}

float stringToFloat(String tmp) {
  char floatbuffer[32];
  tmp.toCharArray(floatbuffer, sizeof(floatbuffer));
  return atof(floatbuffer);
}
```

Project 3 – Building a Twitter Alarm

This project covers the use of ultrasonic range finders and the hypertext transfer protocol (HTTP) POST method. You will learn how to measure distances using the range finder to detect the appearance of an object, and how to send the results to a Twitter account.

Ultrasonic range finders measure the distance to the closest object by sending out sound waves at a frequency that is too high for humans for hear. When the sound wave hits an object, it bounces back. The sensor can calculate how far away an object is by measuring how long it takes for the sound to return. Makers often use this type of sensor when building robots with Arduinos – so that the robot can detect any obstacles in its path and move out of the way.

Unlike infrared devices, which are affected by the amount of ambient light, ultrasonic detectors are less likely to encounter interference and can measure distance more accurately. Even the small, cheap ones have a range of up to 3 or 4 meters. The Parallax Ping))), and the SRF04 and SRF05 devices from Devantech are among the easiest to find and purchase.

Although the Ping))) is slightly different to the SRF04/SRF05, you can work with any of these in this project.

To complete this project, you need:

- An Arduino Uno/Leonardo/Mega 2560/Duemilanove, or compatible board.
- An Arduino Ethernet Shield, or compatible board.
- One LED.
- One 220Ω resistor.
- An ultrasonic range finder, such as the Ping))), SRF04, or SRF05.
- Some wire, and a breadboard or basic soldering equipment.

Although not shown on the connection diagrams, and as with all of the other projects in this book, connect the LED to digital pin 2 through the 220Ω resistor.

First, you will write the parts of the sketch that send messages to Twitter.

Introducing ThingTweet from ThingSpeak

Twitter's application programming interface (API) runs over HTTP, like a webpage. However, instead of sending back hypertext markup language (HTML), it returns data in extensible markup language (XML) format.

Unfortunately, Twitter's security model is difficult to support on a platform with as few resources as the Arduino. Communicating over hypertext transfer protocol secure (HTTPS) and using OAuth, which authenticates users when they login to post tweets, is possible but would require a large amount of work. Most examples of tweeting from an Arduino, including this one, use a service that acts as an intermediary between your sketch and Twitter.

ThingTweet is a Twitter app that acts as a kind of proxy server for simple devices. You can send your tweets to ThingTweet and their system posts the status update messages to Twitter for you. Sending information this way is much simpler than dealing with Twitter's API directly.

To use ThingTweet, you first need to sign up for a free account and authorize it to connect to your Twitter account:

1. Create a free account at **ThingSpeak** by signing up at *https://thingspeak.com/users/ sign_up*
2. On the main **navigation bar**, click **Apps**.
3. Click **ThingTweet**, and then click **Link Twitter Account**.
4. Enter your Twitter username and password, then click **Authorize App**.
5. Click **Back to ThingTweet**.
6. Make a note of the **API Key** associated with the Twitter account.

> **Caution: Since ThingTweet is a Twitter app, you do not need to send your Twitter username and password to it from the Arduino. However, someone could intercept your API key and then post tweets to your account. If this happens, you can change the API key from the ThingSpeak.com website.**

To send tweets from an Arduino sketch, you need to open a connection to the ThingTweet server and make an HTTP request using the POST method.

Start a new sketch in the Arduino IDE and paste in the following code, or type it carefully.

```
#include <SPI.h>
#include <Ethernet.h>

const byte mac[] = { 0x00, 0xC2, 0xA2, 0xE6, 0x3D, 0x57 };
const byte LED = 2;
const char wServer[] = "api.thingspeak.com";
const char wFile[] = "/apps/thingtweet/1/statuses/update";

EthernetClient myClient;

void setup() {
  //D53 on an Arduino Mega must be an output.
  pinMode(53, OUTPUT);

  pinMode(LED, OUTPUT);

  Serial.begin(9600);
  while (!Serial);

  Serial.print("Establishing network connection... ");

  if (Ethernet.begin((uint8_t*)mac) == 0) {
    Serial.println("FAILED!");
    while (true) {
```

```
      digitalWrite(LED, HIGH);
      delay(500);
      digitalWrite(LED, LOW);
      delay(500);
    }
  }
  Serial.println("OK!");
}

void loop() {
}

void skipHeader() {
  char lc;
  while (myClient.connected()) {
    if (myClient.available()) {
      char nc = myClient.read();
      if ((lc == 10) && (nc == 13)) {
        while (myClient.available() == 0);
        myClient.read();
        break;
      }
      else
        lc = nc;
    }
  }
}
```

You start with a basic sketch that connects to the network and defines two constants that describe the web server the sketch is going to connect to.

The server name for ThingTweet is *api.thingspeak.com* and, to send a tweet, you make an HTTP request for the page */apps/thingtweet/1/statuses/update*.

You will need the API Key that you received when you linked ThingTweet to your Twitter account. If you do not have it:

1. Sign in to **ThingSpeak** at *www.thingspeak.com*.
2. On the main **navigation bar**, click **Apps**.
3. Click **ThingTweet**.
4. Make a note of the **API Key** associated with the Twitter account.

Add the API Key to the Arduino sketch as a global array of characters:

```
const char API_Key[] = "THE16DIGITAPIKEY";
```

And declare a new function in the sketch:

```
boolean tweet(String message) {
  boolean result = true;
  digitalWrite(LED, HIGH);
  Serial.println("Connecting to ThingTweet... ");
  if (myClient.connect(wServer, 80) != 1) {
    Serial.println("FAILED!");
```

```
    return false;
  }
  Serial.println("OK!");
  myClient.stop();
  digitalWrite(LED, LOW);
  return result;
}
```

tweet() is not complete yet – you will add code to make an HTTP POST to ThingTweet in the next section.

At the end of the sketch's *setup()* function, add a call to *tweet()*. This shows whether tweets can be sent successfully. For example:

```
tweet("Twitter alarm is online!");
```

Making an HTTP POST Request

In previous projects, you have used the HTTP method GET to request information from a web server. This project uses the POST method. Making a POST is still a request from a client that the server is expected to respond to. However, the POST method tells the server that the client is sending additional information after the usual HTTP request header.

This is the way information is sent to a web server when you submit forms on webpages, and it is also the way that you need to send tweets to ThingTweet.

A valid POST request to ThingTweet looks like this:

```
POST /apps/thingtweet/1/statuses/update HTTP/1.0[crlf]
Host: api.thingspeak.com[crlf]
Connection: close[crlf]
Content-Type: application/x-www-form-urlencoded[crlf]
Content-Length: 72[crlf]
[crlf]
api_key=THE16DIGITAPIKEY&status=This is your message to post on Twitter.
```

Content-Length specifies the number of bytes in the data.

The *Content-Type* field specifies how the data (the string starting with "api_key=") is encoded.

When using the URL encoding scheme, you send key/value pairs as character strings separated by an ampersand. Keys are like variable names, and to work with another system you will have to use the keys that it tells you to. The value of a key/value pair can be any kind of data provided that it is encoded in the correct way. To avoid interfering with the HTTP request, spaces in values should be replaced with a '+', and other non-alphanumeric characters in the data should be replaced by a percent sign followed by the hexadecimal number of the character.

Sending spaces, exclamation marks, and periods as un-encoded characters is usually fine, but they are often encoded anyway. The string "This is your message to post on Twitter!" can be encoded as This+is+your+message+to+post+on+Twitter%21.

If you plan to send a wide variety of non-alphanumeric characters, or do not know what type of characters you will be sending, then you should implement a function to replace non-alphanumeric characters using URL encoding. This project does not do that because the messages are hard-coded into the sketch and contain only basic characters.

The POST request above sends two values to ThingTweet:

1. *api_key* is required so that ThingTweet knows which Twitter account to send the information to.
2. *status* is required by Twitter. It contains the text of the status update.

Aside from *api_key*, ThingTweet passes the key/value pairs on to Twitter so you can add any of the fields from the status update method in Twitter's API:

Field	Description
status	The text of your status update. Usually limited to 140 characters.
in_reply_to_status_id	The ID of an existing status update that this update is in reply to.
lat	Specifies the geographic latitude that this tweet refers to.
long	Specifies the geographic longitude that this tweet refers to.
place_id	A geocode of a particular place in the world.
display_coordinates	Whether or not to put a pin where the tweet has been sent from. This should be either *true* or *false*.
trim_user	Whether or not to display the full author's details when this tweet is displayed in a timeline. This should be either *true* or *false*.
include_entities	Specifies whether the tweet includes additional metadata in an entities node. This should be either *true* or *false*.

You can also include special commands such as direct messaging and retweets. These commands are usually sent at the start of the status update message and, although not used in this project, a list of these can be found at *https://support.twitter.com/articles/14020-twitter-for-sms-basic-features*

Extend the *tweet()* function to send a POST request by adding the following the code before the call to *myClient.stop()*:

```
myClient.print("POST ");
myClient.print(wFile);
myClient.println(" HTTP/1.0");
myClient.print("Host: ");
myClient.println(wServer);
myClient.println("Connection: close");
myClient.println("Content-Type: application/x-www-form-urlencoded");
myClient.print("Content-Length: ");
myClient.println(message.length()+32);
myClient.println();

//POST some data
myClient.print("api_key=");
myClient.print(API_Key);
myClient.print("&status=");
myClient.println(message);
```

The *Content-Length* is 32 plus the number of characters in the message string. 32 refers to the number of characters used in the first part of the data, *api_key=THE16DIGITAPIKEY&status=.*

After the server receives the number of bytes of data indicated by the *Content-Length* field, it processes your request and returns an HTTP response header followed by single value to indicate whether the action succeeded.

To process the server's response:

1. Check that the HTTP status code of the response is 200. For more information, see Reacting to HTTP Status Codes on page 43.

2. Skip the remainder of the HTTP header and read the value sent by ThingTweet, which will usually be only one character.

3. Return *true* if the value is the ASCII character '1' (49), and *false* if it is anything else.

The code for the *tweet()* function should now look similar to this:

```
boolean tweet(String message) {
  boolean result = true;

  digitalWrite(LED, HIGH);
  Serial.print("Connecting to ThingTweet... ");
  if (myClient.connect(wServer, 80) != 1) {
    Serial.println("FAILED!");
    return false;
  }
  Serial.println("OK!");

  myClient.print("POST ");
```

```
myClient.print(wFile);
myClient.println(" HTTP/1.0");
myClient.print("Host: ");
myClient.println(wServer);
myClient.println("Connection: close");
myClient.println("Content-Type: application/x-www-form-urlencoded");
myClient.print("Content-Length: ");
myClient.println(message.length()+32);
myClient.println();

//POST some data
myClient.print("api_key=");
myClient.print(API_Key);
myClient.print("&status=");
myClient.println(message);

char rc = 0;
skipHeader();
while(myClient.connected()) {
  if (myClient.available() > 0) {
    rc = myClient.read();
    break;
  }
}

if (rc == '1')
  result = true;
else
  result = false;

myClient.stop();
digitalWrite(LED, LOW);
return result;
}
```

This function currently ignores the HTTP status code, which is not a recommended course of action, but checking only the ThingTweet return code is usually sufficient while testing.

Measuring Distance with Ultrasonic Range Finders

Connect your ultrasonic range finder to your Arduino as shown in the diagrams below. When using the Ethernet Shield, you can connect the wires into the shield's headers and they will pass through to the Arduino.

Depending on the ultrasonic range finder that you have, you may need to solder wires or a right-angled pin strip to the bottom row of holes on the device. Consult the datasheet for your ultrasonic device before continuing.

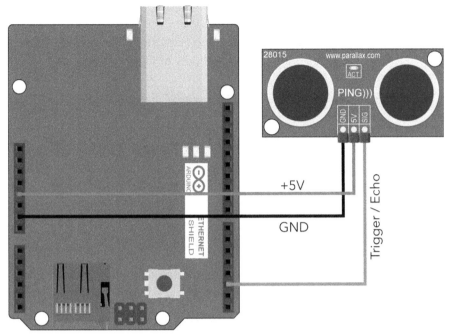

Figure 9. The Ethernet Shield connected to a Ping)))

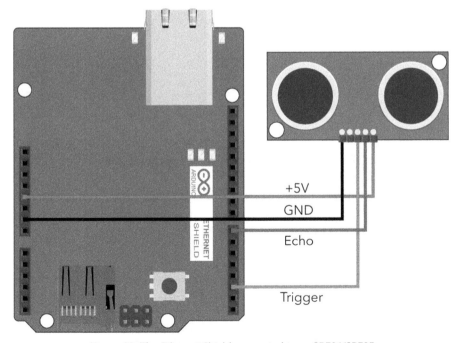

Figure 10. The Ethernet Shield connected to an SRF04/SRF05

When the mode pin of the SRF05 is left unconnected, the device operates in the same way as the SRF04. On the SRF04, this pin would also be unconnected.

Reading from the Ping))) is slightly different to reading from the SRF04 and SRF05. However, the example code in this sketch is designed to work with any of the three devices.

Declare two constants at the top of the sketch: *US_Trigger* and *US_Echo*:

```
const byte US_Trigger = 3;
const byte US_Echo = 7;
```

Ultrasonic range finders typically require a short trigger pulse to trigger them – this is assigned to digital pin 3 on the Arduino.

The time that it takes for the signal to bounce back to the device is measured and returned as a pulse on the echo line, which is assigned to the Arduino's digital pin 7.

The Parallax Ping))) uses trigger and echo signals on the same pin. In this sketch, you can set *US_Trigger* and *US_Echo* to use the same digital pin in the declaration of the constants.

Add this function to the sketch:

```
long getDistance() {
  pinMode(US_Trigger, OUTPUT);
  digitalWrite(US_Trigger, LOW);
  delayMicroseconds(2);
  digitalWrite(US_Trigger, HIGH);
  delayMicroseconds(2);
  digitalWrite(US_Trigger, LOW);

  pinMode(US_Echo, INPUT);
  long duration = pulseIn(US_Echo, HIGH);
  return (duration / 58);
}
```

The calls to *pinMode()* are to support a Ping))). If an SRF04 or SRF05 is connected then two calls to *pinMode()* can be done from the sketch's *setup()* function instead.

After sending a high pulse, the function uses the Arduino library function *pulseIn()* to fetch the time it takes for the ultrasonic signal to return to the range finder. To convert this duration to a distance in centimeters, divide the value by 58.

Completing the Project

Now that the sketch includes a function for reading the distance to the nearest object using the ultrasonic range finder, and a function for posting status updates to a Twitter account, you can combine them so that the sketch tweets when it detects objects in range.

Add these variable declarations to the sketch:

```
const byte MAX_RANGE = 150;
boolean Detected_Object;
```

Depending on the size of your environment and the range of the ultrasonic sensor you are using, the device might detect the walls of the room. To avoid sending tweets unnecessarily, the *MAX_RANGE* constant specifies how close an object has to be to count as detected.

Detected_Object is used to ensure that the same object is not detected continuously, generating hundreds of tweets.

If you have not done so already, add a call to *tweet()* in the sketch's *setup()* function to send a message stating that the Arduino is online and ready. Since *tweet()* returns *false* if the message is not sent successfully to ThingTweet, you can continually retry sending the message using a while loop:

```
String message = "I am online, and watching... (";
message.concat(random(0, 0xFFFFul));
message.concat(")");
while (tweet(message) == false) {
  delay(5000);
}
```

Twitter sometimes filters messages that it considers might be duplicates or spam. Concatenating a random number to the message adds something different to each tweet so that Twitter is more likely to accept them.

In the sketch's *loop()* function, call *getDistance()* to see if an object is detected and to find its distance.

- If there is an object, and *Detected_Object* is currently *false*, create a message.
- If there is an object but *Detected_Object* is *true* then no action is taken – since the tweet for this reading has already been sent.
- If there is no object and *Detected_Object* is *true* then this indicates that there was an object in front of the sensor but it has now been removed. Create a different message for this.

Finally, check if a message string was created, if it was, tweet it by passing the message string as an argument to *tweet()*.

Source Code

The complete sketch for this project is shown below. Remember to replace *API_Key* with the ThingTweet API key for your own Twitter account.

```
#include <SPI.h>
#include <Ethernet.h>

const byte mac[] = { 0x00, 0xC2, 0xA2, 0xE6, 0x3D, 0x57 };
const byte LED = 2;
const byte US_Trigger = 3;
const byte US_Echo = 7;
const byte MAX_RANGE = 250;

boolean Detected_Object;
EthernetClient myClient;

const char wServer[] = "api.thingspeak.com";
const char wFile[] = "/apps/thingtweet/1/statuses/update";
const char API_Key[] = "THE16DIGITAPIKEY";

void setup() {
  //D53 on an Arduino Mega must be an output.
  pinMode(53, OUTPUT);

  pinMode(LED, OUTPUT);

  Serial.begin(9600);
  while (!Serial);

  Serial.print("Establishing network connection... ");

  if (Ethernet.begin((uint8_t*)mac) == 0) {
    Serial.println("FAILED!");
    while (true) {
      digitalWrite(LED, HIGH);
      delay(500);
      digitalWrite(LED, LOW);
      delay(500);
    }
  }
  Serial.println("OK!");

  randomSeed(analogRead(2));
  String message = "I am online, and watching... (";
  message.concat(random(0, 0xFFFFul));
  message.concat(")");
  while (tweet(message) == false) {
    delay(5000);
  }
  Detected_Object = false;

  // Delay for 15s to allow time for humans to move out of the way.
  delay(15000);
}

void loop() {
  String message;

  long distance = getDistance();
  if (distance < MAX_RANGE) {
      if (!Detected_Object) {
          message = "I have detected an unknown object at ";
          message.concat(distance);
          message.concat("cm  (";
```

```
          message.concat(random(0, 0xFFFFul));
          message.concat(")");
          Detected_Object = true;
      }
    }
    else {
      if (Detected_Object) {
        message = "Phew! It's gone! (";
        message.concat(random(0, 0xFFFFul));
        message.concat(")");
        Detected_Object = false;
      }
    }

    if (message.length() > 0) {
      Serial.println(message);
      while(tweet(message) == false) {
        delay(1000);
      }
    }

    delay(500);
}

long getDistance() {
  pinMode(US_Trigger, OUTPUT);
  digitalWrite(US_Trigger, LOW);
  delayMicroseconds(2);
  digitalWrite(US_Trigger, HIGH);
  delayMicroseconds(2);
  digitalWrite(US_Trigger, LOW);

  pinMode(US_Echo, INPUT);
  long duration = pulseIn(US_Echo, HIGH);
  return (duration / 58);
}

boolean tweet(String message) {
  boolean result = true;

  digitalWrite(LED, HIGH);
  Serial.print("Connecting to ThingTweet... ");
  if (myClient.connect(wServer, 80) != 1) {
    Serial.println("FAILED!");
    return false;
  }
  Serial.println("OK!");

  myClient.print("POST ");
  myClient.print(wFile);
  myClient.println(" HTTP/1.0");
  myClient.print("Host: ");
  myClient.println(wServer);
  myClient.println("Connection: close");
  myClient.println("Content-Type: application/x-www-form-urlencoded");
  myClient.print("Content-Length: ");
  myClient.println(message.length()+32);
  myClient.println();

  //POST some data
  myClient.print("api_key=");
  myClient.print(API_Key);
  myClient.print("&status=");
  myClient.println(message);

  char rc = 0;
  skipHeader();
  while(myClient.connected()) {
    if (myClient.available() > 0) {
```

```
      rc = myClient.read();
      break;
    }
  }

  if (rc == '1')
    result = true;
  else
    result = false;

  myClient.stop();
  digitalWrite(LED, LOW);
  return result;
}

void skipHeader() {
  char lc;
  while (myClient.connected()) {
    if (myClient.available()) {
      char nc = myClient.read();
      if ((lc == 10) && (nc == 13)) {
        while (myClient.available() == 0);
        myClient.read();
        break;
      }
      else
        lc = nc;
    }
  }
}
```

Arduino as a Web Server

Servers wait for incoming connections from clients, process requests for information, and then send the information or an error message indicating why the server could not complete the request. This is the core logic that all servers implement – whether they are web servers, database servers, or even part of an online game – regardless of how complicated the task of processing the request might be.

In Arduino as a Web Client on page 29, you can see how to open connections using the *EthernetClient* class from the Arduino's Ethernet library, and how to work with the hypertext transfer protocol (HTTP) as a web client. In this chapter you will be using the *EthernetServer* class to accept incoming connections and work with the HTTP protocol as a web server: processing HTTP request messages and sending back HTTP response messages.

In Project 4 – Setting up a Basic Web Server you will also learn about using a static IP address, port forwarding, and dynamic DNS, so that other machines are able to find and connect to your Arduino.

In This Chapter

Project 4 – Setting up a Basic Web Server

The Ethernet Shield is not designed so that the Arduino can host large, professional websites – the Arduino's small amount of memory and relatively slow speed make that very difficult. It is designed so that projects can make their information and features available to other computers, through an interface that is well-known and flexible: the web browser.

In this project you will write a sketch that:

1. Connects to the network and waits for incoming connections from clients.
2. Extracts the requested file name from clients' HTTP requests.
3. Reads the value of an analog sensor, and returns this as part of a webpage.

Many of the details of HTTP and the roles of clients and servers is covered earlier in this book. If you have not done so already, you should read Arduino as a Web Client on page 29 for an explanation of the way request and response messages are handled in HTTP.

To complete this project, you need:

- An Arduino Uno/Leonardo/Mega 2560/Duemilanove, or compatible board.
- An Arduino Ethernet Shield, or compatible board.
- One light-dependent resistor (LDR, or photocell/phototransistor).
- One 10KΩ resistor.
- Some wire, and a breadboard or basic soldering equipment.

Using a Static IP Address

So far, you have connected your Arduino to the network using dynamic host configuration protocol (DHCP). With DHCP, your network router assigns the Arduino an IP address when it connects. In Establishing a Network Connection on page 12, you can see how this is done and how the IP address assigned by the router can be sent to the serial port and used by another machine to ping the device.

However, when building servers, DHCP has one large drawback. When the Arduino reconnects to the network, the router may not assign the same IP address to the Arduino that it used previously. This can make it difficult for clients to find the Arduino and make a connection to it.

The sketch can demand the same IP address each time it connects by specifying the IP address as an array of four bytes and passing this array into the call to *Ethernet.begin()*.

```
const byte ip[] = { 192, 168, 0, 99 };
```

Each of the four parts in an IP address is a number in the range 0 through 255.

The first three parts should be the same as other devices on your network, and this is usually 192.168.0 for most home networks. The final part must be unique to this device and, as most routers begin assigning IP addresses from 192.168.0.2 and work upwards, it is usually safest to pick a large number.

When called with an IP address, *Ethernet.begin()* does not return a value. To check whether the connection has been established, you can make the call to *Ethernet.begin()* and then verify that *Ethernet.localIP()*, *Ethernet.gatewayIP()*, and *Ethernet.subnetMask()* return sensible values.

The starting point for this project is a sketch that makes a network connection using the IP address declared near the top of the file, just after the library inclusions and the declaration of the media access control (MAC) address. It then creates an instance of the *EthernetServer* class, specifying 80 as the port it will listen on. Finally, it calls the *EthernetServer* class method *begin()* to start the server.

```
#include <SPI.h>
#include <Ethernet.h>

const byte mac[] = { 0x00, 0xC3, 0xA2, 0xE6, 0x3D, 0x57 };
const byte ip[] = { 192, 168, 0, 99 };

EthernetServer myServer(80);

void setup() {
  //D53 on an Arduino Mega must be an output.
  pinMode(53, OUTPUT);

  Serial.begin(9600);
  while (!Serial);

  Serial.println("Establishing network connection...");

  Ethernet.begin((uint8_t*)mac, (uint8_t*)ip);

  Serial.print("IP Address: ");
  Serial.println(Ethernet.localIP());

  Serial.print("Default Gateway: ");
  Serial.println(Ethernet.gatewayIP());

  Serial.print("Subnet Mask: ");
  Serial.println(Ethernet.subnetMask());

  Serial.print("DNS Server: ");
  Serial.println(Ethernet.dnsServerIP());

  myServer.begin();
}

void loop() {
}
```

To check that everything is working, ping the device using the instructions in Testing the Connection on page 15. The sketch won't respond to web requests until you add code to accept connections in the sketch's *loop()* function.

Introducing Port Forwarding and Dynamic DNS

Web clients make connections to web servers over transmission control protocol (TCP) port 80, and the IP address tells the client where it can find the web server.

However, when your computers connect to the Internet through a router, your Internet service provider (ISP) assigns an additional IP address to the router. To the outside world, all devices on your local network appear to have the same IP address – the one set by the ISP. The local network addresses that you have been working with so far are not used by clients connecting from the Internet.

This creates two problems: how can your router know which machine on your network is supposed to respond to web requests, and how do clients on the Internet find your router?

Port forwarding is a configuration setting that you can use to tell your router which device on your local area network (LAN) should receive the connection from the outside world. Unfortunately, the exact method of setting this up is different for each router.

As a general guide:

1. Login to your router's administration panel. For most home routers, this is usually done by visiting the URL http://192.168.0.1 in a web browser on your PC.
2. Look for an option or page that allows you to control inbound connections. This may be named Port Forwarding, Firewall Rules, Services, or something similar.
3. Create a rule that says the HTTP service (TCP:80) is allowed, and should be sent to the LAN server/machine 192.168.0.99 (the IP address set in the Arduino sketch).

When you visit a website, you use its domain name (for example, google.com) as part of the web address. Your web browser looks up the domain name to find the associated IP address, using the domain name system (DNS). Clients make connections to servers using these IP addresses.

But it is unlikely that your ISP has allocated a fixed (or static) IP address to your router. Instead, the IP address will change every time your router connects to the Internet. If the connection is dropped, or the router restarts, this address will change.

Dynamic DNS (DDNS) is a method of automatically updating DNS records when the router receives a new IP address from an ISP. This means that any web clients trying to find your server using a domain name will always be given the up-to-date IP address.

Although there are many DDNS services out there, they all tend to work the same way: when your router connects to the Internet then either it or a machine on the LAN contacts the DDNS servers and tells them the new IP address. You can usually choose whether to use a domain name given to you by the DDNS service, or buy your own domain name and use that.

Accepting Connections

The method *available()* in the *EthernetServer* class returns an instance of *EthernetClient* if a client is waiting to the connect to the web server. If no clients are waiting, *available()* returns a result that evaluates to *false*.

This should be called in the sketch's *loop()* function so that the server is able to keep responding to connections while the Arduino has power. Change the sketch's *loop()* function to match the following code sample:

```
void loop() {
  EthernetClient client = myServer.available();
  if (client) {
    Serial.println("Incoming connection...");
    client.stop();
  }
}
```

This *loop()* accepts connection requests from clients, but then closes the connection without sending any data to the client. It only sends a message to the serial port so that you can see when incoming connections are detected. These messages are viewable from the serial port monitor in the Arduino IDE.

Using a web browser on your PC, or a tool such as the W3C's markup validation service at *validator.w3.org*, you should be able to connect to the server and receive the HTTP error 500.

To respond to the web client, you must send an HTTP response header followed by data (if the request was successful). For example:

```
HTTP/1.0 200 OK[crlf]
Content-Type: text/plain[crlf]
Connection: close[crlf]
[crlf]
Connection received OK.
```

The sketch in this project does not check which version of HTTP the client wants to use, it always sends back responses using HTTP/1.0. But most modern web browsers are extremely tolerant when it comes to servers returning unexpected, or partial, responses.

For testing purposes, add code to the *loop()* function to send the HTTP response shown above:

```
void loop() {
  EthernetClient client = myServer.available();
  if (client) {
    Serial.println("Incoming connection...");
    client.println("HTTP/1.0 200 OK");
    client.println("Content-Type: text/plain");
    client.println("Connection: close");
    client.println();
```

```
    client.println("Connection received OK.");
    client.stop();
  }
}
```

Machines on the local network can connect using the IP address sent to the serial port during the sketch's *setup()* function. If your network router (and any DNS or DDNS services) is configured correctly then machines connecting across the Internet should use the external IP address of the router. All clients should receive the message "Connection received OK."

Reading from an Analog Sensor

This project uses a light-dependent resistor (LDR, photocell, or photoresistor) connected to the Ethernet Shield's A0 input. When the Ethernet Shield is connected to the Arduino, A0 will pass through to the Arduino's analog input 0. If you do not have an LDR, you can use the ultrasonic range finder and *getDistance()* function from Project 3 – Building a Twitter Alarm on page 51, or a random number generated using the Arduino function *random()*.

To connect an LDR to the Arduino for use in this project, build a circuit similar to this:

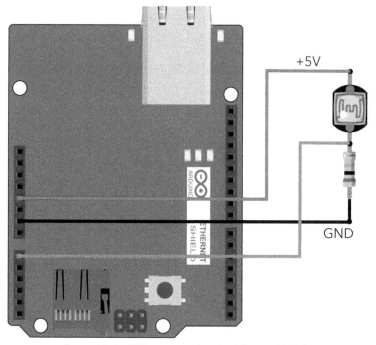

Figure 11. An LDR connected to the Ethernet Shield

The Arduino uses an analog to digital converter (ADC) to measure the voltage level of the input pin. When the LDR is exposed to light, its resistance decreases and so the voltage read by the ADC is high. When light is blocked, the resistance of the LDR increases and so the voltage read by the ADC is lower.

The 10KΩ resistor and LDR form a potential divider, which protects the Arduino from short circuits by ensuring that there is always some resistance on the line.

Add this function to the sketch:

```
int getAnalogReading() {
  return analogRead(0);
}
```

If you are using a different sensor, change the *getAnalogReading()* function to return a value from whatever sensor you are working with.

Returning Webpages

Returning a proper webpage is not very different from returning the test message. The process is:

1. Send back an HTTP response that tells the web browser that the request completed successfully (HTTP status code 200).
2. Send the *Content-Type* header field with the value *text/html*.[1]
3. Send the HTML code for the webpage.

Create a new function in the sketch, name this *sendWebpage()*.

```
void sendWebpage(EthernetClient client) {
}
```

Add these lines to the function's body to send an HTTP response header which indicates that the request was successful and that the browser should expect an HTML file:

```
client.println("HTTP/1.0 200 OK");
client.println("Content-Type: text/html");
client.println("Connection: close");
client.println();
```

---

1. The Content-Type field is discussed in more detail in Understanding MIME and Media Types on page 83.

The remainder of the function sends the HTML page in three parts. To do this:

1. Make calls to *client.print()* to send HTML code until the value of the sensor is needed.

2. Make a call to *client.print()* and pass the value returned by the function *getAnalogReading()*.

3. Make calls to *client.print()* and send the remaining HTML code for the document.

The full source code for *sendWebpage()* is at the end of this project. You can copy it into your sketch now if you are unsure how to write the HTML code.

In the sketch's *loop()* function, replace the calls to *client.println()* that send back the test message with a call to the *sendWebpage()* function. *loop()* should look this:

```
EthernetClient client = myServer.available();
if (client) {
  Serial.println("Incoming connection...");
  sendWebpage(client);
  client.stop();
}
```

On your PC, open your web browser and type *http://192.168.0.99/* (or whatever IP address you set in the sketch) into the address bar. If you use the HTML code from this project's Source Code on page 75 then the page will look something like Figure 12.

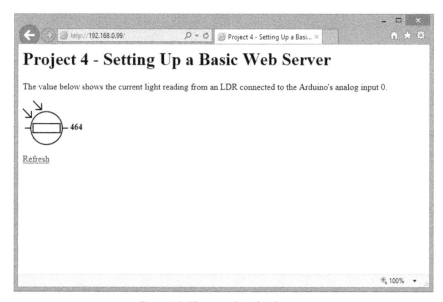

Figure 12. The completed webpage

> **Tip: The HTML page sent by this sketch includes a reference to an image of an LDR from a different website – the image itself is not sent to the browser by the Arduino.** Project 5 – Building a More Advanced Web Server on page 77 **shows how files such as images can be loaded from the SD card and sent to web clients.**

Examining the Request URI

Even if your project is only going to return one page, it is usually sensible to examine the contents of the request line (in particular, the request URI) that the client sent to your server.

"Request URI" is just another term that refers to the name of the file or resource that the web browser is asking for. In a typical HTTP request, the request URI ends at the next space character. However, if the browser sends a query string then the file name part of the URI ends at a question mark – for more information about query strings, see Understanding Query Strings and POST Data on page 92.

The client could be asking for an icon or other file that your project does not have. On those occasions, you should return an HTTP error message. But at the moment, the sketch sends back the webpage displaying the reading from the LDR, regardless of what the client asked for.

To ensure that the webpage is only returned when the client requests / or /index.html you can:

1. Check that the request method is "GET" followed by a space.

2. Read the request URI into a *String* object, until you find a space or a question mark.

3. Check if the string is equal to / or /index.html. Return an HTTP error *404 – file not found* if it is not. Call s*endWebpage()* if it is.

Replace the sketch's *loop()* function with this code:

```
void loop() {
  EthernetClient client = myServer.available();
  if (client) {
    Serial.println("Incoming connection...");

    while (client.connected()) {
      if (client.available() >= 4)
      break;
    }

    char cMethod[5] = {'G', 'E', 'T', ' ', 0};
    char buf1[5] = {0, 0, 0, 0, 0};
    client.read((uint8_t*)buf1, 4);
    if (strcmp(buf1, cMethod) == 0) {
      String cFile = "";
      while (client.connected()) {
        if (client.available() > 0) {
          char tmp = client.read();
          if ((tmpc != ' ') && (tmpc != '?'))
```

```
            cFile.concat(tmp);
          else
            break;

          if (cFile.length() > 200) {
            client.println("HTTP/1.0 414 Request Too Long");
            client.println("Connection:close");
            client.println();
            cFile = "";
            break;
          }
        }
      }

      if (cFile != "") {
        if (
            (cFile.compareTo("/") == 0) ||
            (cFile.compareTo("/index.html") == 0)
          ) {
            sendWebpage(client);
          }
        else {
          client.println("HTTP/1.0 404 File Not Found");
          client.println("Connection:close");
          client.println();
        }
      }
    }
    else {
      client.println("HTTP/1.0 501 Not Implemented");
      client.println("Connection:close");
      client.println();
    }

    client.stop();
  }
}
```

The loop waits until at least four bytes of the HTTP request have been received. The code then checks whether the HTTP request uses the GET method: if it is a GET request then the first four characters are "GET" followed by a space.

The buffer array (*buf1*) and the character array containing the sequence of characters to compare to the buffer (*cMethod*), both have an extra zero at the end. This is because the Arduino function *strcmp()* expects null-terminated strings.

If the two character arrays are not equal, then the request method is not "GET" and the server returns an HTTP error *501 – not implemented*.

But if the two arrays are equal, the sketch proceeds to read the file path and file name from the request. Since whitespace should be encoded with a percent sign in HTTP requests, the routine can read the file name until it finds a space or a question mark, without any spaces in the actual file name causing a problem.

The maximum length of a web address is around 2000 characters – but this can vary between different web browsers and web servers. 2000 characters is too long for Arduino-based servers to process, and so the complete sketch shown at the end of this project returns an HTTP error *414 – request too long* if the file name exceeds 200 characters.

When the while loop exits – hopefully as a result of reading the full file path and file name from the request – the code checks whether the requested file is */* or */index.html* (this string comparison is case sensitive). If that is true then it calls *sendWebpage()* to send the HTML data for the page. If the request asked for any other file then the code returns a HTTP error *404 — file not found*.

Source Code

This is the complete source code for the basic web server sketch.

```
#include <SPI.h>
#include <Ethernet.h>

const byte mac[] = { 0x00, 0xC3, 0xA2, 0xE6, 0x3D, 0x57 };
const byte ip[] = { 192, 168, 0, 99 };

EthernetServer myServer(80);

void setup() {
  //D53 on an Arduino Mega must be an output.
  pinMode(53, OUTPUT);

  Serial.begin(9600);
  while (!Serial);

  Serial.println("Establishing network connection...");

  Ethernet.begin((uint8_t*)mac, (uint8_t*)ip);

  Serial.print("IP Address: ");
  Serial.println(Ethernet.localIP());

  Serial.print("Default Gateway: ");
  Serial.println(Ethernet.gatewayIP());

  Serial.print("Subnet Mask: ");
  Serial.println(Ethernet.subnetMask());

  Serial.print("DNS Server: ");
  Serial.println(Ethernet.dnsServerIP());

  myServer.begin();
}

void loop() {
  EthernetClient client = myServer.available();
  if (client) {
    Serial.println("Incoming connection...");

    while (client.connected()) {
      if (client.available() >= 4)
      break;
    }

    char cMethod[5] = {'G', 'E', 'T', ' ', 0};
    char buf1[5] = {0, 0, 0, 0, 0};
    client.read((uint8_t*)buf1, 4);
    if (strcmp(buf1, cMethod) == 0) {
```

```
        String cFile = "";
        while (client.connected()) {
          if (client.available() > 0) {
            char tmp = client.read();
            if ((tmp != ' ') && (tmp != '?'))
              cFile.concat(tmp);
            else
              break;

            if (cFile.length() > 200) {
              client.println("HTTP/1.0 414 Request Too Long");
              client.println("Connection:close");
              client.println();
              cFile = "";
              break;
            }
          }
        }

        if (cFile != "") {
          if (
              (cFile.compareTo("/") == 0) ||
              (cFile.compareTo("/index.html") == 0)
              ) {
                sendWebpage(client);
              }
          else {
            client.println("HTTP/1.0 404 File Not Found");
            client.println("Connection:close");
            client.println();
          }
        }
      }
      else {
        client.println("HTTP/1.0 501 Not Implemented");
        client.println("Connection:close");
        client.println();
      }

      client.stop();
  }
}

int getAnalogReading() {
  return analogRead(0);
}

void sendWebpage(EthernetClient client) {
  client.println("HTTP/1.0 200 OK");
  client.println("Content-Type: text/html");
  client.println("Connection:close");
  client.println();
  client.print("<html>");
  client.print("<head>");
  client.print("<title>Project 4 - Setting Up a Basic Web Server</title>");
  client.print("</head>");
  client.print("<body>");
  client.print("<h1>Project 4 - Setting Up a Basic Web Server</h1>");
  client.print("<p>The value below shows the current light reading from ");
  client.print("an LDR connected to the Arduino's analog input 0.</p>");
  client.print("<table cellpadding=0 cellspacing=0 border=0>");
  client.print("<tr>");
  client.print("<td valign='middle'><img src='http://www.gcse.com/nldr.gif' /></td>");
  client.print("<td> </td>");
  client.print("<td valign='middle'><b style='padding-top: 15px; display: block'>");

  client.print(getAnalogReading());

  client.print("</b></td>");
```

```
    client.print("</tr>");
    client.print("</table>");
    client.print("<p><a href='/'>Refresh</a></p>");
    client.print("</body>");
    client.print("</html>");
}
```

Project 5 – Building a More Advanced Web Server

In Project 4 you can see how to build a web server that responds to requests over HTTP and sends back hypertext markup language (HTML) code that can be displayed by any web browser. This project expands on that functionality by building a web server that serves files from the SD card, and processes webpage "templates" instead of hard-coding each line of the HTML document into the sketch.

When working with the Arduino Ethernet Shield and SD card libraries, you frequently have to reduce the amount of error checking and robustness of the sketch in order to fit it into the memory available on the Arduino. For this reason, this project does not make several of the checks that are made in earlier projects.

Other techniques that you can use to reduce the amount of memory consumed by the sketch include:

- Using a single, global array of characters for file name processing, instead of *String* objects.

- Using pointers to avoid duplication of *File* objects when passing arguments to functions.

- Passing strings to functions as pointers to *const __FlashStringHelper* objects, and using the macro *F()* so that the compiler places the strings in Flash memory, not SRAM.

This project does not explain the code used to work with microSD cards. For information about the SD card socket built onto the Arduino Ethernet Shield and using the SD.h library, refer to Using SD Cards on page 18.

The starting point for this project is a sketch that initializes the SD card and makes a connection to the network using a static IP address as described in Using a Static IP Address on page 66. The code in the sketch's *loop()* function waits for an incoming HTTP request and attempts to extract the file path and file name that the web browser requests.

```
#include <SPI.h>
#include <Ethernet.h>
#include <SD.h>
```

```
const byte mac[] = { 0x00, 0xC0, 0xA2, 0xE6, 0x3D, 0x54 };
const byte ip[] = { 192, 168, 0, 99 };

EthernetServer myServer(80);
EthernetClient client;
char fname[100];

void setup() {
  //D53 on the Arduino Mega must be an output.
  pinMode(53, OUTPUT);

  SD.begin(4);

  Ethernet.begin((uint8_t*)mac, (uint8_t*)ip);
  myServer.begin();
}

void sendError(int code, const __FlashStringHelper *message) {
  client.print(F("HTTP/1.0 "));
  client.print(code);
  client.print(" ");
  client.println(message);
  client.println(F("Connection: close"));
  client.println();
  client.println(message);
}

void readFileRequest() {
  byte c = 0;
  char tmpc;
  while (client.connected()) {
    if (client.available() > 0) {
      tmpc = client.read();
      if ((tmp != ' ') && (tmp != '?'))
        fname[c++] = tmpc;
      else {
        fname[c] = 0;
        break;
      }

      if (c > 100) {
        sendError(414, F("Request Too Long"));
        fname[0] = 0;
        break;
      }
    }
  }
}

boolean sendDirectoryList(File *di) {
    return false;
}

boolean sendFile(File *fi) {
    return false;
}

void loop() {
  client = myServer.available();
  if (client) {
      while (client.connected()) {
        if (client.available() >= 4) {
          client.read();
          client.read();
          client.read();
          client.read();
          break;
        }
```

```
      }
      if (client.connected()) {
        readFileRequest();

        // Process the file request here.

      }
      client.stop();
  }
}
```

When an incoming HTTP request is received in the sketch's *loop()* function, the while loop waits until the client sends the first four characters of the request. This should be the word "GET" followed by a space. Due to difficulties fitting this project on an Arduino, assume that the client is using the GET request method. If it uses another method or sends an invalid HTTP request, then the problem is detected either when the sketch tries to read to the file name, or when it tries to find a file matching that name on the SD card. The Arduino might send back the wrong error message, but it will send back an error message.

The sketch also includes a function for sending HTTP error messages, *sendError()*, and declares two incomplete functions: *sendDirectoryList()* and s*endFile()*. You will complete *sendDirectoryList()* and *sendFile()* in this project but, for now, have a look at the function *sendError()* in the code above. Its declaration is

```
void sendError(int code, const __FlashStringHelper *message)
```

The function has two parameters:

code is the HTTP status code of the error.

message is the description of the error.

The reason for using the type *const __FlashStringHelper* is to reduce the amount of SRAM consumed when working with strings. To pass a string argument into this function, you can no longer use:

```
sendError(500, "Internal Server Error")
```

Instead, you must pass the string into the *F()* macro, which stores the string in Flash memory:

```
sendError(500, F("Internal Server Error"))
```

Browsing Directories

Web servers normally disable directory browsing. However, in this project you will complete the functions *sendDirectoryList()* and *sendFile()* to send webpages that list the files and folders on the SD card, and allows visitors to click on files to download them.

To implement this, there are four checks to run in the sketch's *loop()* method:

1. If the requested file is /, open the root directory of the SD card and call *sendDirectoryList()*.

2. If the requested file is not / but it is a valid directory on the SD card, open the directory and call *sendDirectoryList()*.

3. If the file exists but it is not a directory, open the file and call *sendFile()*.

4. If the requested file does not exist, call *sendError(404, F("File Not Found"))* to send back an error message.

Insert the following code into the sketch, replacing the comment "*// Process the file request here*":

```
if ( (fname[0] == '/') && (fname[1] == 0) ) {
  File di = SD.open(fname);
  if (!sendDirectoryList(&di))
    sendError(500, F("Internal Server Error"));
  di.close();
}
else {
  if (SD.exists(fname)) {
    File tmp = SD.open(fname);
    if (tmp.isDirectory())
      sendDirectoryList(&tmp);
    else
      sendFile(&tmp);
    tmp.close();
  }
  else
    sendError(404, F("File Not Found"));
}
```

In the function *sendDirectoryList()*, add code to output a simple HTML page that displays a list of the contents of a directory. This function comprises the following steps:

1. Check that the directory was opened successfully.

2. *Rewind* the directory using the method *rewindDirectory()*. This ensures that the buffers used by the SD library are cleared. If you do not do this, sometimes the directory cannot be fully listed.

3. Send a response header that tells the client that the request is successful, using the *Content-Type* field to tell the client to expect an HTML page.

4. Send the first part of the HTML page to the client.

5. For each item in the directory, send HTML that shows the item's name and a link to it.

6. Send the last part of the HTML page to the client.

For example:

```
boolean sendDirectoryList(File *di) {
  if (*di) {
    di->rewindDirectory();

    client.println(F("HTTP/1.0 200 OK"));
    client.println(F("Content-Type: text/html"));
    client.println(F("Connection: close"));
    client.println();
    client.println(F("<html>"));
    client.print(F("<head><title>"));
    client.print(fname);
    client.println(F("</title></head>"));
    client.println(F("<body>"));
    client.print(F("<h1>Index of "));
    client.print(fname);
    client.println(F("</h1>"));
    client.println(F("<table cellpadding=2 cellspacing=2 border=0>"));

    File lsf;
    while ((lsf = di->openNextFile())) {
      client.println(F("<tr>"));
      client.print(F("<td>"));
      if (lsf.isDirectory())
        client.print(F("[dir]"));
      else
        client.print(lsf.size());
      client.print(F("</td>"));
      client.print(F("<td>"));
      client.print(F("<a href='"));
      client.print(fname);
      if (fname[1] != 0)
        client.print(F("/"));
      client.print(lsf.name());
      client.print(F("'>"));
      client.print(lsf.name());
      client.print(F("</a>"));
      client.print(F("</td>"));
      client.println(F("</tr>"));
      lsf.close();
    }

    client.println(F("</table>"));
```

```
      client.println(F("</body>"));
      client.println(F("</html>"));

      return true;
   }
   else
      return false;
}
```

Note the use of -> because the argument passed into *sendDirectoryList()* is a pointer to an instance of the *File* class, it is not an actual object.

To read the contents of the directory, this routine uses the same method as shown in Reading from SD Cards on page 21. The file path and file name of the current directory are added to each link in the HTML output, so that the links contain the full file path for each item.

To complete the function *sendFile()*, the process is much shorter:

1. Check that the file was opened successfully.

2. Send a response header that tells the client that the request is successful, using the *Content-Type* field to tell the client to expect a file.[1]

3. Read each byte, one at a time, from the SD card and send it to the client.

The following code is an example of how to do this:

```
boolean sendFile(File *fi) {
   if (*fi) {
      client.println(F("HTTP/1.0 200 OK"));
      client.println(F("Content-Type: application/octet-stream"));
      client.println(F("Connection: close"));
      client.println();
      while (fi->available()) {
         client.write(fi->read());
      }
      return true;
   }
   else
      return false;
}
```

> **Tip: Remember that the Arduino SD library only supports file names in the format 8:3. If a client directly requests a file name that does not conform to the 8:3 format then it will not be found on the SD card. Using 8:3 is also advantageous as you do not usually have to worry about encoding and decoding "special" characters using the URL encoding scheme.**

---

1. The media type application/octet-stream informs the browser to expect a binary file that it can process however it feels is best.

Understanding MIME and Media Types

Web browsers use the *Content-Type* field in an HTTP response header to decide what to do with the file – this serves the same purpose as the file extension on Microsoft Windows.

The values of the *Content-Type* field are often called *MIME types* because they are derived from the multipurpose Internet mail extensions (MIME) standard, which is used to allow email messages to contain attachments and different types of text. In the HTTP protocol specification they are called *media types*.

Some of the most common media types used on the web are:

File Extension	Media Type	Description
.HTM	text/html	Specifies that the file is an HTML document.
.GIF	image/gif	Specifies that the file is an image in the graphics interchange format (GIF).
.PNG	image/png	Specifies that the file is an image in the portable network graphics (PNG) format.
.JPG	image/jpeg	Specifies that the file is an image in the format created by the Joint Photographic Experts Group.
.CSS	text/css	Specifies that the file is a cascading style sheet (CSS).
.TXT	text/plain	Specifies that the file is a plain text document.

The *sendFile()* function in the sketch currently sends all files with the media type *application/octet-stream*. This will often work fine, as the web browser will interpret the file and decide how to proceed. However, if possible, you should send the correct *Content-Type* field for the file that you are sending.

To ensure that HTML documents are sent as *text/html* and C source files are sent as *text/plain*: change the line in *sendFile()* that reads *client.println("Content-Type: application/octet-stream ");* and replace it with code that checks the file extension in the array *fname*.

```
client.print(F("Content-Type: "));
byte sl = strlen(fname);
if (sl > 4) {
  if ((fname[sl-4]=='.') && (fname[sl-3]=='H') && (fname[sl-2]=='T') && (fname[sl-1]=='M'))
    client.println(F("text/html"));
  if ((fname[sl-4]=='.') && (fname[sl-3]=='C') && (fname[sl-2]=='S') && (fname[sl-1]=='S'))
    client.println(F("text/css"));
  else
    client.println(F("application/octet-stream"));
```

```
} else if (sl > 2) {
  if ((fname[sl-2]=='.') && ( (fname[sl-1]=='H') || (fname[sl-1]=='C')) )
    client.println(F("text/plain"));
  else
    client.println(F("application/octet-stream"));
}
else
  client.println(F("application/octet-stream"));
```

You should do this for each of the file types that your project is going to host. However, if you run out of space in the sketch, it may be acceptable to send important files with the correct *Content-Type* and others with *application/octet-stream*.

At this point, the web server sketch serves HTML and CSS files in a way that web browsers can understand. If you decide to use a CSS file in a webpage, you should save the style sheet in the 8:3 file name format. For example: *NORMAL.CSS*. You can add references to this style sheet in your HTML files by using a *link* element in the *head* element of each webpage:

```
<link href="/NORMAL.CSS" rel="stylesheet" type="text/css">
```

Using Template Webpages

In Project 4 – Setting up a Basic Web Server on page 65, you create a sketch that sends a webpage to the client. This webpage is dynamic, meaning that it changes every time it is accessed, because the value of the light-dependent resistor (LDR) changes.

Sending the page involves mixing calls to the *EthernetClient* class method *print()* – sometimes printing fragments of HTML code and sometimes numeric values. However, hard-coding the HTML into the sketch makes it inconvenient to design and update the webpages used by your project.

Some programming languages – such as active server pages (ASP), java server pages (JSP), and PHP – use code tags inserted into the HTML files. When the server loads the file, it replaces these tags with actual values before it sends the data to the client. Sometimes called "templates", in modern website programming the same files can be used to control the output of multiple different webpages. Usually, the server accepts values in the query string of an HTTP request and uses these values to decide what data to insert into the template. Query strings are examined in more detail in Project 6 – Controlling Digital Outputs from the Web on page 91.

To complete this project, extend the *sendFile()* method so that it:

1. Sends files with the extension .ASP ("Arduino server pages") using the media type *text/html.*
2. Uses a different while loop to read and process .ASP files on the SD card.
3. Detects code tags and replaces them with values.

To send the correct media type, you can add another else if clause to the block of if statements that decide the value of the *Content-Type* HTTP response field. Add this code after the call to *client.println("text/css"):*

```
else if ((fname[sl-4]=='.') && (fname[sl-3]=='A') && (fname[sl-2]=='S') && (fname[sl-1]=='P'))
  client.println("text/html");
```

The code tags used in this project are formed by two percent signs followed by a single character. For example, %%r.

Save the HTML code below to a file named *TEST.ASP* in the root directory of your SD card.

```
<html>
<head>
<title>Project 5 - Building a More Advanced Web Server</title>
</head>
<body>
<h1>Project 5 - Building a More Advanced Web Server</h1>
<p>Demonstrates pre-processing a file and creating template webpages.</p>
<p>Random number: %%r.</p>
<p>Not a valid code tag: %%e.</p>
<p>Just a percent sign: %</p>
<p><a href='/TEST.ASP'>Refresh</a></p>
</body>
</html>
```

In order for the sketch to replace these tags with values, add the following code to the *sendFile()* function in your sketch. Place this routine after the call to *client.println()* that sends the blank line marking the end of the HTTP response header fields.

```
if (sl > 4) {
  if ((fname[sl-4]=='.') && (fname[sl-3]=='A') && (fname[sl-2]=='S') && (fname[sl-1]=='P')) {
    char tmpc;
    boolean found_mark1 = false;
    boolean found_mark2 = false;
    while (fi->available()) {
      tmpc = fi->read();
      if ((tmpc == '%') && (!found_mark1))
        found_mark1 = true;
      else if ((tmpc == '%') && (found_mark1))
        found_mark2 = true;
      else {
        if ((found_mark1) && (found_mark2)) {
          switch (tmpc) {
            case 'r':
              client.print(random());
              break;
```

```
          }
        }
        else if ((found_mark1) && (!found_mark2)) {
          client.print('%');
          client.write(tmpc);
        }
        else
          client.write(tmpc);
        found_mark1 = found_mark2 = false;
      }
    }
    return true;
  }
}
```

If the requested file has the extension .ASP, then this routine reads characters from the file until it finds two percent signs next to each other. When that happens, the next character that is read from the file is processed in a switch statement to determine what the sketch writes into the HTML page instead of the code tag. Once the routine completes, it returns true so that the remainder of the *sendFile()* function does not run.

If the sketch finds one percent sign, but does not find a second immediately after it, it sends a percent sign in addition to the character that has just been read from the file. This is so that individual percent signs, not being used as code tags, are sent correctly.

When accessed from a web browser, the file *TEST.ASP* looks like Figure 13.

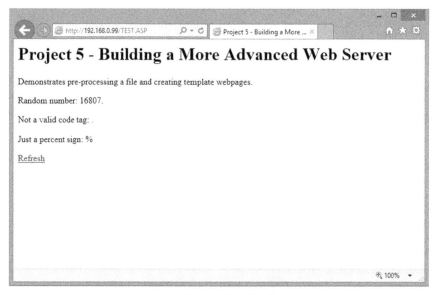

Figure 13. The completed webpage

In cases where the character following the marker %% is an 'r', this code sends a random number using the Arduino's function *random()*. You can certainly extend this logic to insert variables, or values obtained by reading analog sensors.

For example, to extend the sketch to support a code tag %%p, which inserts the level of light measured by a photocell:

1. Connect a light-dependent resistor (LDR, or photocell) using the instructions in Reading from an Analog Sensor on page 70.

2. Copy the function *getAnalogReading()* from Project 4 – Setting up a Basic Web Server on page 65 and paste it into this sketch.

3. Add the following statements after the break statement in the function *sendFile()*:

```
case 'p':
    client.print(getAnalogReading());
    break;
```

Concerning Multiple Connections

Modern web servers handle requests from thousands of clients simultaneously. However, even though the Arduino Ethernet Shield can support up to four clients connected at the same time, there are further limitations.

The Ethernet library that comes with the Arduino IDE only supports one connection per port. While you can accept one connection on port 80 and another connection on port 81, you cannot keep two connections on port 80 independent.

In certain applications, it is often acceptable to allocate different ports for different clients. To do so:

1. Create an instance of the *EthernetServer* class for each port (up to four).

2. In the sketch's *loop()* function, check each *EthernetServer* instance in turn.

3. Get a different instance of the *EthernetClient* class from each instance of *EthernetServer*.

4. Process each request one by one.

If you do try to declare multiple instances of *EthernetServer* using the same port, you will find that writing to one instance of the *EthernetClient* class causes the data to be sent to all of the instances sharing that port.

There is another limitation to be aware of before doing this: the lack of multitasking or threading on the Arduino platform. Without these features, a loop like the one below will still prevent any other clients from connecting when the server is sending a large file.

```
while(fi->available()) {
  client.write(fi->read());
}
```

To solve this problem, you need to change how HTTP requests are processed – only reading and sending a few bytes from each file before dropping back to the sketch's *loop()* function in order to do the same on another connection. This means storing the state of up to four requests, and it is not a trivial task to do all of this without exhausting the Arduino's limited resources.

Source Code

The complete source code for this project is shown below.

```
#include <SPI.h>
#include <Ethernet.h>
#include <SD.h>

const byte mac[] = { 0x00, 0xC0, 0xA2, 0xE6, 0x3D, 0x54 };
const byte ip[] = { 192, 168, 0, 99 };

EthernetServer myServer(80);
EthernetClient client;
char fname[100];

void setup() {
  //D53 on the Arduino Mega must be an output.
  pinMode(53, OUTPUT);
  SD.begin(4);
  Ethernet.begin((uint8_t*)mac, (uint8_t*)ip);
  myServer.begin();
}

void sendError(int code, const __FlashStringHelper *message) {
  client.print(F("HTTP/1.0 "));
  client.print(code);
  client.print(" ");
  client.println(message);
  client.println(F("Connection: close"));
  client.println();
  client.println(message);
}

void readFileRequest() {
  byte c = 0;
  char tmpc;
  while (client.connected()) {
    if (client.available() > 0) {
      tmpc = client.read();
      if ((tmpc != ' ') && (tmpc != '?'))
        fname[c++] = tmpc;
      else {
        fname[c] = 0;
        break;
      }
      if (c > 100) {
```

```
            sendError(414, F("Request Too Long"));
            fname[0] = 0;
            break;
        }
      }
    }
}

boolean sendDirectoryList(File *di) {
  if (*di) {
    di->rewindDirectory();

    client.println(F("HTTP/1.0 200 OK"));
    client.println(F("Content-Type: text/html"));
    client.println(F("Connection: close"));
    client.println();
    client.println(F("<html>"));
    client.print(F("<head><title>"));
    client.print(fname);
    client.println(F("</title></head>"));
    client.println(F("<body>"));
    client.print(F("<h1>Index of "));
    client.print(fname);
    client.println(F("</h1>"));

    client.println(F("<table cellpadding=2 cellspacing=2 border=0>"));

    File lsf;
    while ((lsf = di->openNextFile())) {
      client.println(F("<tr>"));
      client.print(F("<td>"));
      if (lsf.isDirectory())
        client.print(F("[dir]"));
      else
        client.print(lsf.size());
      client.print(F("</td>"));
      client.print(F("<td>"));
      client.print(F("<a href='"));
      client.print(fname);
      if (fname[1] != 0)
        client.print(F("/"));
      client.print(lsf.name());
      client.print(F("'>"));
      client.print(lsf.name());
      client.print(F("</a>"));
      client.print(F("</td>"));
      client.println(F("</tr>"));
      lsf.close();
    }

    client.println(F("</table>"));
    client.println(F("</body>"));
    client.println(F("</html>"));

    return true;
  }
  else
    return false;
}

boolean sendFile(File *fi) {
  if (*fi) {
    client.println(F("HTTP/1.0 200 OK"));

    client.print(F("Content-Type: "));
    byte sl = strlen(fname);
    if (sl > 4) {
      if ((fname[sl-4]=='.') && (fname[sl-3]=='H') && (fname[sl-2]=='T') && (fname[sl-1]=='M'))
        client.println(F("text/html"));
```

```
          else if ((fname[sl-4]=='.') && (fname[sl-3]=='C') && (fname[sl-2]=='S') && (fname[sl-1]=='S'))
            client.println(F("text/css"));
          else if ((fname[sl-4]=='.') && (fname[sl-3]=='A') && (fname[sl-2]=='S') && (fname[sl-1]=='P'))
            client.println(F("text/html"));
          else
            client.println(F("application/octet-stream"));
        } else if (sl > 2) {
          if ((fname[sl-2]=='.') && ( (fname[sl-1]=='H') || (fname[sl-1]=='C')) )
            client.println(F("text/plain"));
          else
            client.println(F("application/octet-stream"));
        }
        else
          client.println(F("application/octet-stream"));

        client.println(F("Connection: close"));
        client.println();

        if (sl > 4) {
          if ((fname[sl-4]=='.') && (fname[sl-3]=='A') && (fname[sl-2]=='S') && (fname[sl-1]=='P')) {
            char tmpc;
            boolean found_mark1 = false;
            boolean found_mark2 = false;
            while (fi->available()) {
              tmpc = fi->read();
              if ((tmpc == '%') && (!found_mark1))
                found_mark1 = true;
              else if ((tmpc == '%') && (found_mark1))
                found_mark2 = true;
              else {
                if ((found_mark1) && (found_mark2)) {
                  switch (tmpc) {
                    case 'r':
                      client.print(random());
                      break;
                  }
                }
                else if ((found_mark1) && (!found_mark2)) {
                  client.print('%');
                  client.write(tmpc);
                }
                else
                  client.write(tmpc);
                found_mark1 = found_mark2 = false;
              }
            }
            return true;
          }
        }

        while (fi->available()) {
          client.write(fi->read());
        }
        return true;
      }
    else
      return false;
}

void loop() {
  client = myServer.available();

  if (client) {
      while (client.connected()) {
        if (client.available() >= 4) {
          client.read();
          client.read();
          client.read();
          client.read();
```

```
        break;
      }
    }

    if (client.connected()) {
      readFileRequest();

      if (
        (fname[0] == '/') &&
        (fname[1] == 0)
        )
      {
        File di = SD.open(fname);
        if (!sendDirectoryList(&di))
          sendError(500, F("Internal Server Error"));
        di.close();
      }
      else {
        if (SD.exists(fname)) {
          File tmp = SD.open(fname);
          if (tmp.isDirectory())
            sendDirectoryList(&tmp);
          else
            sendFile(&tmp);
          tmp.close();
        }
        else
          sendError(404, F("File Not Found"));
      }

    }
    client.stop();
  }
}
```

Project 6 – Controlling Digital Outputs from the Web

This project is centered on passing information from a web browser to the Arduino and using this information to control digital outputs and devices, such as light-emitting diodes (LEDs), servos or motors. In this project you will see how to return JavaScript object notation (JSON) instead of webpages, and build user interfaces (UIs) in hypertext markup language (HTML) and JavaScript so that the web browser does not need to refresh the page in order to communicate with the web server.

In effect, you are going to build an application programming interface (API) that works over hypertext transfer protocol (HTTP).

To complete this project you will need:

- An Arduino Uno/Leonardo/Mega 2560/Duemilanove, or compatible board.
- An Arduino Ethernet Shield, or compatible board.
- A microSD card, formatted for use with the Ethernet Shield[1].
- At least one (up to four) RC servo motors, the three-pin type.
- A 5V power source. For example, a 9V battery and a 7805 voltage regulator.
- Some wire, and a breadboard or basic soldering equipment.

If you have worked through Project 4 and Project 5, you have already seen one way that web browsers send information to web servers – the name (and path) of the file that they are requesting. You have also seen how to interpret this file name – by comparing it to predefined strings, examining the file extensions, or passing the file name directly to the SD library in order to send a file back to the web browser.

Project 3 introduced POST requests in HTTP. This is another way that web browsers send information to web servers. From the point of view of a web server, POST requests are interpreted by checking whether the request method in the HTTP request is POST instead of GET. If it is, extract the value from the *Content-Length* header field and then skip the remaining parts of the header. Finally, read the number of bytes specified by Content-Length: this is the data that the web browser sent in key/value pairs.

Understanding Query Strings and POST Data

In Making an HTTP POST Request on page 55, you can see how data can be encoded into key/value pairs and sent to a web server using the POST method. In a GET request, this same style of encoding can be used to append the information to the name of the file in an HTTP request. For example:

```
GET /myfile.txt?key1=value1&key2=value2 HTTP/1.0[crlf]
Connection: close[crlf]
[crlf]
```

A web browser sending this request is asking for the file *myfile.txt*, and it is sending two parameters that the server can use help it process the request. Each parameter is made up of a key and a value which are separated by an equals sign. Parameters are separated by ampersands, and the data string is separated from the file name by a question mark.

---

1. See Formatting and Initializing SD Cards on page 18

This string *?key1=value1&key2=value2* is called a query string.

While query strings and POST data are very simple to work with on most systems, they are often inconvenient in Arduino projects due to the amount of memory it takes to store and process them.

Introducing JavaScript, jQuery, and JSON

JavaScript is a client-side scripting language that you can use to add interactivity and functions to webpages, without the web browser needing to reload the page every time the user clicks on something. The code actually runs in the web browser on the visitor's machine and manipulates the contents of the loaded webpage. jQuery is a library that is written in JavaScript, and which simplifies the language and provides additional methods to make writing client-side scripts a little easier. You can include it in your webpages by downloading the .js file from www.jquery.com; saving it to the microSD card you are using with the Ethernet Shield; and then adding a script element to the webpages, specifying the file name as the source.[1]

Or you can link directly to a distribution of jQuery by adding the following script block in the page's *head* element:

```
<script src="http://code.jquery.com/jquery-1.11.1.min.js"></script>
```

It is beyond the scope of this book to provide a guide to programming in JavaScript and jQuery. However, in this project, you will modify the web server from Project 5 – Building a More Advanced Web Server on page 77. so that it correctly returns JavaScript files, and receives commands from the client-side script running in the visitor's web browser. These commands will control four servos connected to the Arduino.

JSON represents objects and collections of objects in a structured text format – typically organized into key/value pairs but each value can also be another collection of key/value pairs. The JavaScript interpreter in web browsers can create a data object directly from JSON, and this has helped make it extremely popular for use in web APIs.

A JSON object looks like this:

```
{
  "result": "ok",
  "values": [1, 2, 3, 4, 5, 6, 7, 8, 9]
}
```

When evaluated by the JSON parser, this data creates an object with two properties: *result* (which contains the text string "ok"), and *values* (an array of numbers, 1–9).

---

1. Remember to rename the file so that it follows the 8:3 file name format.

jQuery includes the method *getJSON()* which you can use to make a web request to the Arduino without reloading the current webpage. *getJSON()* expects a response from the web server in JSON format, and will automatically parse this response into a JavaScript object.

You will also need the jQuery Rotate plugin from code.google.com/p/jqueryrotate/. Download the file *jQueryRotate.js* from the website and save it to the root of the SD card as *ROTATE.JS*.

To ensure that JavaScript (.JS) files are sent with the correct *Content-Type*, you will need to modify the Arduino web server that you created in Project 5. In the function *sendFile()*, before the line

```
} else if (sl > 2) {
```

Add the following code to the if statement:

```
} else if (sl > 3) {
  if ((fname[sl-3]=='.') && (fname[sl-2]=='J') && (fname[sl-1]=='S'))
    client.println(F("application/javascript"));
  else
    client.println(F("application/octet-stream"));
}
```

The example HTML webpage used in this book includes images in the portable network graphics (PNG) format. To ensure that these images are sent with the correct *Content-Type*, after the lines:

```
else if ((fname[sl-4]=='.') && (fname[sl-3]=='A') && (fname[sl-2]=='S') && (fname[sl-1]=='P'))
  client.println(F("text/html"));
```

Add the following code:

```
else if ((fname[sl-4]=='.') && (fname[sl-3]=='P') && (fname[sl-2]=='N') && (fname[sl-1]=='G'))
  client.println(F("image/png"));
```

Connecting the Servos

Connect the servo motors to the Arduino as shown in Figure 14. If you are only using one servo, you can connect it directly to the Arduino's 5V output (Figure 15). However, as with all types of motor, servos can draw significant amounts of current and the Arduino is not a suitable power source if you are using more than one or two servos.

A 7805 voltage regulator provides a stable 5V power source when fed from 9V battery, or similar input supply. Remember to connect the Arduino's ground to this circuit.

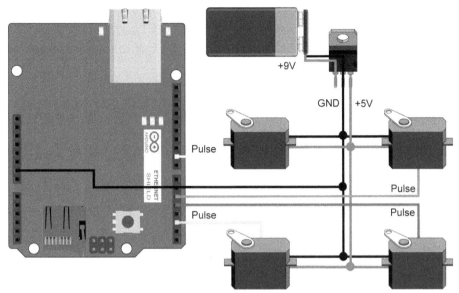

Figure 14. Connecting four servos to the Ethernet Shield

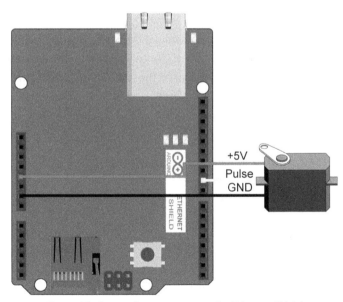

Figure 15. Connecting one servo to the Ethernet Shield

You control servos using pulses, and the easiest way to do this is to connect the servos to the Arduino (through the Ethernet Shield) on pins that are marked "PWM". Since the Ethernet Shield uses digital pin 10, connect the servos on pins 9, 6, 5, and 3.

Standard servos allow the rotation of the shaft to be set between 0 and 180 degrees, with 90 as the center point. This sketch should work with all standard RC (hobby) servo motors that connect via three pins – 5V, pulse, ground. However, you should check the datasheet for the parts that you have, to ensure that you are operating the component within its capabilities and have wired it up correctly.

The Arduino IDE comes with a library for controlling servos – *Servo.h*. Include this library at the top of the sketch:

```
#include <Servo.h>
```

Add the following definitions near to the top of the Arduino sketch:

```
Servo servo1, servo2, servo3, servo4;
byte Angle1 = 90;
byte Angle2 = 90;
byte Angle3 = 90;
byte Angle4 = 90;
```

This creates four instances of the *Servo* class and these control the servo motors connected to the Arduino. The other four lines declare four bytes that hold the current angle of rotation of each servo. At the start of the sketch, all four servos are at zero degrees rotation.

To prepare each servo for use, call the method *attach()* from each instance of the *Servo* class. Then use the method *write()* to move the servo into its center position – it may have moved while connecting the wires, or could be out of position if the sketch has been run previously. Add the following lines to the *setup()* function in your sketch:

```
servo1.attach(9);
servo2.attach(6);
servo3.attach(5);
servo4.attach(3);
servo1.write(90);
servo2.write(90);
servo3.write(90);
servo4.write(90);
```

Building the User Interface

Create an HTML document that includes four textboxes. Set the ID attribute of these textboxes to *angle1*, *angle2*, *angle3* and *angle4* respectively. These textboxes are to contain the angle of rotation of each of the four servos connected to the Arduino.

Above each textbox, add an image that will represent the current position of the servo. For example, you could use the image of a knob, such as a volume knob on a guitar amplifier. Set the

ID attribute of these images to *knob1*, *knob2*, *knob3* and *knob4* respectively. The JavaScript code will rotate these knobs to match the position of the servos.

In the webpage's *head* element, add a script reference to jQuery and jQuery Rotate.

The webpage shown in Figure 16 is made from the following HTML code:

```
<html>
<head>
  <title>Project 6 — Controlling Digital Outputs from the Web</title>
  <script src="http://ajax.googleapis.com/ajax/libs/jquery/1.11.1/jquery.min.js"></script>
  <script src="/ROTATE.JS"></script>
  <style type="text/css">
    body { background-color: #D0D2D3; }
    h1 { text-align: center; }
    input { width: 50px; border: 1px solid black; margin-right: 30px; }
    p { text-align: center; }
    #panel { display: block;
             width: 340px; height: 100px;
             margin: 30px auto 10px auto; padding: 0 0 0 30px;
           }
    .knob { display: inline-block; width: 50px; background: url(BACK.PNG) top left no-repeat;
            margin-right: 30px;
          }
  </style>
</head>
<body>
  <h1>Project 6 — Controlling Digital Outputs from the Web</h1>
  <div id="panel">
    <div class="knob"><img src="KNOB.PNG" alt="" id="knob1" /></div>
    <div class="knob"><img src="KNOB.PNG" alt="" id="knob2" /></div>
    <div class="knob"><img src="KNOB.PNG" alt="" id="knob3" /></div>
    <div class="knob"><img src="KNOB.PNG" alt="" id="knob4" /></div>
    <br/>
    <input type="text" value="0" id="angle1" />
    <input type="text" value="0" id="angle2" />
    <input type="text" value="0" id="angle3" />
    <input type="text" value="0" id="angle4" />
  </div>
  <p>Enter a number between 0 and 180 in a box and press Return/Enter.</p>
</body>
</html>
```

The knobs in Figure 16 are created using two portable network graphics (PNG) files – one for the knob itself, and another that is displayed underneath and provides a little background shadow.

Save this webpage to the root of the SD card that you are using with the Arduino, with a file name in the 8:3 format, such as *SERVOS.HTM*.

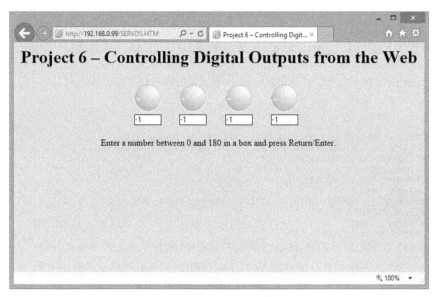

Figure 16. The completed UI

The next step is to add some JavaScript control to the textboxes using jQuery. In the webpage's *head* element, add the following script block:

```
<script language="javascript">
  // defines a function that allows only numbers and some basic cursor
  // control keys to be pressed when one of the textboxes is selected.
  function checkBox (e) {
    if (
        ($.inArray(e.keyCode, [46, 8, 9, 27, 110, 190]) !== -1) ||
        (e.keyCode == 65 && e.ctrlKey === true) ||
        (e.keyCode >= 35 && e.keyCode <= 39)
      ) {
        return;
    }
    if (e.keyCode == 13) {
      e.preventDefault();
      var angle = parseInt(e.currentTarget.value);
      if (angle < 0)
        angle = 0;
      else if (angle > 180)
        angle = 180;
      e.currentTarget.value = angle;
      e.currentTarget.blur();
      sendRotation(e.currentTarget.id, angle);
    }
    else if (
            (e.shiftKey || (e.keyCode < 48 || e.keyCode > 57)) &&
            (e.keyCode < 96 || e.keyCode > 105)
          ) {
              e.preventDefault();
    }
  }

  function sendRotation(box, angle) {
  }
```

```
  // add the handlers to the four textboxes
  $(document).ready(function() {
    $("#angle1").keydown(checkBox);
    $("#angle2").keydown(checkBox);
    $("#angle3").keydown(checkBox);
    $("#angle4").keydown(checkBox);
  });
</script>
```

The JavaScript function *checkBox()* accepts a jQuery event object (which contains details about the "event" that occurs when the user presses a key when one of the textboxes has the focus) and uses this information to block any non-numeric characters from the textbox – with the exception of certain special key presses such as the *Home* and *End* keys.

When the user presses the *Enter* key (13), *checkBox()* converts the string that is in the textbox to an integer number, and passes this number into the function, *sendRotation()*. Currently, *sendRotation()* does nothing, but will be used to send the command to the Arduino.

Before *checkBox()* can respond to user key presses, you must add it as a handler for the keydown event on all four of the textboxes. This is done from jQuery's *ready()* function so that the web browser only attempts to add the handlers once the full document has been received from the web server, ensuring that the textboxes are first included in the document object model (DOM).

Receiving Commands on the Arduino

The JavaScript code in the completed UI makes HTTP requests for one of two pages:

Webpage	Description
/gs	Returns the current position of all four servos connected to the Arduino.
/ss	Sets the rotation of a particular servo.

When the JavaScript code makes a request for the file */gs*, the sketch will currently return a file not found error message, because the file is not on the SD card. You want to change this behavior.

In the sketch's *loop()* function, find the lines that check whether the file exists on the SD card. The final three lines of this code send the error message:

```
}
else
  sendError(404, F("File Not Found"));
```

Before the curly brace and else statement, add the following code to process requests for */gs*:

```
else if ( (fname[0]=='/') && (fname[1]=='g') && (fname[2] == 's') && (fname[3]==0) ) {
  client.println(F("HTTP/1.0 200 OK"));
  client.println(F("Content-Type: application/json"));
  client.println(F("Cache-Control: no-cache, no-store, must-revalidate"));
  client.println(F("Pragma: no-cache"));
  client.println(F("Expires: 0"));
  client.println(F("Connection: close"));
  client.println();
  client.print(F("{\"result\":\"ok\", \"values\":["));
  client.print(Angle1);
  client.print(F(","));
  client.print(Angle2);
  client.print(F(","));
  client.print(Angle3);
  client.print(F(","));
  client.print(Angle4);
  client.print(F("]}"));
}
```

This code causes the sketch to send the values of the four "Angle" variables as JSON data, {"result":"ok", "values":[0,0,0,0]}, when the web browser requests /gs.

Processing requests for /ss is a little more involved. When complete, your JavaScript code will append the number of the servo and the angle of rotation to the file name. This is a little simpler for the Arduino sketch to process than a full query string.

The first line of the HTTP request will look a little like this:

```
GET /ss1270 HTTP/1.1
```

In the sketch's *loop()* function, add a clause to the if statement that checks the requested file. This clause checks:

1. That the file name is at least three characters long.
2. That the first three characters of the requested file name are /ss.
3. That the file name contains characters after the /ss, and whether or not to return an error message in JSON format.

The sketch must then take the first character after the /ss as the servo number, and any remaining characters (there will be at least one, and a maximum of three, in a valid request) as the angle of rotation. These characters will need to be converted to a byte value using the function *atoi()*.

In the sketch's *loop()* function, find the lines that check whether the file exists on the SD card. The final three lines of this code send the error message:

```
}
else
  sendError(404, F("File Not Found"));
```

Before the curly brace and *else* statement, add the following code:

```
else if ((strlen(fname) >= 3) && (fname[0]=='/') && (fname[1]=='s') && (fname[2]=='s') ) {
  if (strlen(fname)==3) {
    client.println(F("HTTP/1.0 200 OK"));
    client.println(F("Content-Type: application/json"));
    client.println(F("Cache-Control: no-cache, no-store, must-revalidate"));
    client.println(F("Pragma: no-cache"));
    client.println(F("Expires: 0"));
    client.println(F("Connection: close"));
    client.println();
    client.println(F("{\"result\":\"error\", \"message\":\"Invalid command\"}"));
  }
  else {
    char buffer[4];
    byte i;
    for (i=4; i < strlen(fname); i++)
      buffer[i-4] = fname[i];
    buffer[i-4] = 0;
    byte angle = atoi(buffer);
    switch (fname[3]) {
      case '0':
        Angle1 = angle;
        servo1.write(180 - angle);
        break;
      case '1':
        Angle2 = angle;
        servo2.write(180 - angle);
        break;
      case '2':
        Angle3 = angle;
        servo3.write(180 - angle);
        break;
      case '3':
        Angle4 = angle;
        servo4.write(180 - angle);
        break;
    }
    client.print(F("{\"result\":\"ok\", \"values\":["));
    client.print(Angle1);
    client.print(F(","));
    client.print(Angle2);
    client.print(F(","));
    client.print(Angle3);
    client.print(F(","));
    client.print(Angle4);
    client.print(F("]}"));
  }
}
```

If the file name is only three characters long then the web client may have made a request for /ss but did not include the servo number and angle of rotation. In these circumstances, the sketch returns a successful HTTP response but the JSON data tells the JavaScript code that there is an error.

If the web client did send all of the necessary information, the code then attempts to convert the characters that represent the angle of rotation into a number. Note that this code does very little error checking.

The switch statement is used to set the "Angle" variables appropriately, depending on which servo is selected, and move the servo using the method *write()* from the selected instance of the *Servo* class.

> **Tip: If you have a servo that turns clockwise, not counterclockwise, change the calls to *write()* so that the angle is not subtracted from 180.**

Finally, the */ss* command completes its operation by returning the status of the all of the servers, in the same way as a */gs* command.

Sending Commands from JavaScript

Sending commands to the Arduino sketch from the JavaScript in your HTML page is done using the function *getJSON()* to request one of the two special webpages – */ss* or */gs*.

getJSON() only accepts two parameters: the full URL of the page to request (including the protocol, http://), and some code to run when the request completes.

Change the *sendRotation()* function in your webpage so that it looks like this:

```
function sendRotation(box, angle) {
  var servo = 0;
  switch (box) {
    case "angle1":
      servo = 0;
      break;
    case "angle2":
      servo = 1;
      break;
    case "angle3":
      servo = 2;
      break;
    case "angle4":
      servo = 3;
      break;
  }
  $.getJSON("/ss" + servo + angle, function(data) {
    if (data.result == "ok") {
      updateDisplay(data);
    } else {
      alert(data.message);
    }
  });
}
```

When *sendRotation()* is called by the *checkBox()* function, it receives the *box* argument to determine which of the four textboxes the user was typing in. This value is the ID attribute of the textbox, and *sendRotation()* first works out the servo number that the ID refers to.

The argument *angle* is the number that the user has typed into the textbox.

These two pieces of information are appended to the file name that the call to *getJSON()* requests from the Arduino. When that request completes, the function that is written into the call to *getJSON()* is executed. In this case, if the server returned a JSON object with the property *result* set to "ok" then the data is passed into *updateDisplay()*, which updates the contents of the textboxes and rotates the images.

Add *updateDisplay()* to your script block:

```
function updateDisplay(data) {
  $("#angle1").val(data.values[0]);
  $("#angle2").val(data.values[1]);
  $("#angle3").val(data.values[2]);
  $("#angle4").val(data.values[3]);
  $("#knob1").rotate(data.values[0]);
  $("#knob2").rotate(data.values[1]);
  $("#knob3").rotate(data.values[2]);
  $("#knob4").rotate(data.values[3]);
}
```

Now when the user types a number into a textbox and presses *Enter*, *sendRotation()* sends the number they type to the Arduino and the UI is updated to show the most recent positions of all four servos.

However, when the webpage is refreshed, the values in the textboxes return to zero. You can add some code into the jQuery *ready()* function so that when the page is loaded, it fetches the most recent positions of the servos from the Arduino. Do this by making a *getJSON()* call to */gs* from the *ready()* function.

For example:

```
$.ajaxSetup({cache:false});
$.getJSON("/gs", function(data) {
  if (data.result == "ok") {
    updateDisplay(data);
  }
});
```

The use of the *ajaxSetup()* function is discussed in the next section.

Controlling the Cache

The term cache (also, caching) refers to web clients storing a temporary copy of files or information that they download from web servers. If the file is needed again, the client can reload it from its cache and avoid the lengthy process of contacting the server and downloading it again.

However, for dynamic data or pages that change frequently, this can be a problem. For example, if the browser caches a request for */ss145* then the next time you enter 45 into the second textbox, the browser will not contact the Arduino. This means that the value will not be sent to the sketch, the servo won't move, and the browser will use out-of-date values to rotate the images.

There are several HTTP response header fields that can be used to help control how pages and data are cached. In this project, when sending JSON data to the web browser, the sketch sends the fields:

Field	Value	Description
Cache-Control	no-cache, no-store, must-revalidate	An HTTP/1.1 directive that asks browsers not to cache the response.
Pragma	no-cache	An HTTP/1.0 directive that asks browsers not to cache the response.
Expires	0	Informs the browser that this webpage expires immediately.

jQuery's *getJSON()* function caches more than most web browsers. In addition to sending the HTTP header fields, you should turn off this caching in jQuery. Include this statement in the jQuery *ready()* function:

```
$.ajaxSetup({cache:false});
```

With caching disabled, jQuery actually adds unique numbers to the query string of each web request. In the source code for this project, you can see that the function *readFileRequest()* already detects a question mark as the end of the file name, and it ignores query strings.

Without this mechanism, even a simple request for */gs* would include a query string that your sketch needs to account for.

Source Code – SERVOS.HTM

Shown below is the HTML and JavaScript code for the UI.

```
<html>
<head>
  <title>Project 6 — Controlling Digital Outputs from the Web</title>
  <script src="http://code.jquery.com/jquery-1.11.1.min.js"></script>
  <script src="ROTATE.JS"></script>
  <style type="text/css">
    body { background-color: #D0D2D3; }
    h1 { text-align: center; }
    input { width: 50px; border: 1px solid black; margin-right: 30px; }
    p { text-align: center; }
    #panel { display: block;
```

```
                  width: 340px; height: 100px;
                  margin: 30px auto 10px auto; padding: 0 0 0 30px;
            }
      .knob { display: inline-block; width: 50px; background: url(BACK.PNG) top left no-repeat;
            margin-right: 30px;
      }
</style>
<script language="javascript">
    // defines a function that allows only numbers and some basic cursor
    // control keys to be pressed when one of the textboxes is selected.
    function checkBox (e) {
      if (
          ($.inArray(e.keyCode, [46, 8, 9, 27, 110, 190]) !== -1) ||
          (e.keyCode == 65 && e.ctrlKey === true) ||
          (e.keyCode >= 35 && e.keyCode <= 39)
         ) {
          return;
      }
      if (e.keyCode == 13) {
        e.preventDefault();

        var angle = parseInt(e.currentTarget.value) || 0;
        if (angle > 180)
          angle = 180;
        e.currentTarget.value = angle;
        e.currentTarget.blur();
        sendRotation(e.currentTarget.id, angle);
      }
      else if (
          (e.shiftKey || (e.keyCode < 48 || e.keyCode > 57)) &&
          (e.keyCode < 96 || e.keyCode > 105)
         ) {
          e.preventDefault();
      }
    }

    function sendRotation(box, angle) {
      var servo = 0;
      switch (box) {
        case "angle1":
          servo = 0;
          break;
        case "angle2":
          servo = 1;
          break;
        case "angle3":
          servo = 2;
          break;
        case "angle4":
          servo = 3;
          break;
      }
      $.getJSON("/ss" + servo + angle, function(data) {
        if (data.result == "ok") {
          updateDisplay(data);
        } else {
          alert(data.message);
        }
      });
    }

    function updateDisplay(data) {
      $("#angle1").val(data.values[0]);
      $("#angle2").val(data.values[1]);
      $("#angle3").val(data.values[2]);
      $("#angle4").val(data.values[3]);
      $("#knob1").rotate(data.values[0]);
      $("#knob2").rotate(data.values[1]);
      $("#knob3").rotate(data.values[2]);
```

```
        $("#knob4").rotate(data.values[3]);
      }

      // add the handlers to the four textboxes and load the
      // position of the servos.
      $(document).ready(function() {
        $("#angle1").keydown(checkBox);
        $("#angle2").keydown(checkBox);
        $("#angle3").keydown(checkBox);
        $("#angle4").keydown(checkBox);

        $.ajaxSetup({cache:false});
        $.getJSON("/gs", function(data) {
          if (data.result == "ok") {
            updateDisplay(data);
          }
        });
      });
    </script>
  </head>
  <body>
    <h1>Project 6 — Controlling Digital Outputs from the Web</h1>
    <div id="panel">
      <div class="knob"><img src="KNOB.PNG" alt="" id="knob1" /></div>
      <div class="knob"><img src="KNOB.PNG" alt="" id="knob2" /></div>
      <div class="knob"><img src="KNOB.PNG" alt="" id="knob3" /></div>
      <div class="knob"><img src="KNOB.PNG" alt="" id="knob4" /></div>
      <br/>
      <input type="text" value="0" id="angle1" />
      <input type="text" value="0" id="angle2" />
      <input type="text" value="0" id="angle3" />
      <input type="text" value="0" id="angle4" />
    </div>
    <p>Enter a number between 0 and 180 in a box and press Return/Enter.</p>
  </body>
</html>
```

Source Code - Sketch

This is the complete source code for the Arduino sketch, which serves files from the SD card in addition to processing servo control messages from the UI.

```
#include <SPI.h>
#include <Ethernet.h>
#include <SD.h>
#include <Servo.h>

const byte mac[] = { 0x00, 0xC0, 0xA2, 0xE6, 0x3D, 0x54 };
const byte ip[] = { 192, 168, 0, 99 };

EthernetServer myServer(80);
EthernetClient client;

char fname[100];

Servo servo1, servo2, servo3, servo4;
byte Angle1 = 90;
byte Angle2 = 90;
byte Angle3 = 90;
byte Angle4 = 90;

void setup() {
  //D53 on the Arduino Mega must be an output.
  pinMode(53, OUTPUT);
```

```
  SD.begin(4);
  servo1.attach(9);
  servo2.attach(6);
  servo3.attach(5);
  servo4.attach(3);
  servo1.write(90);
  servo2.write(90);
  servo3.write(90);
  servo4.write(90);

  Ethernet.begin((uint8_t*)mac, (uint8_t*)ip);
  myServer.begin();
}

void sendError(int code, const __FlashStringHelper *message) {
  client.print(F("HTTP/1.0 "));
  client.print(code);
  client.print(" ");
  client.println(message);
  client.println(F("Connection: close"));
  client.println();
  client.println(message);
}

void readFileRequest() {
  byte c = 0;
  char tmpc;
  while (client.connected()) {
    if (client.available() > 0) {
      tmpc = client.read();
      if ((tmpc != ' ') && (tmpc != '?'))
        fname[c++] = tmpc;
      else {
        fname[c] = 0;
        break;
      }
      if (c > 100) {
        sendError(414, F("Request Too Long"));
        fname[0] = 0;
        break;
      }
    }
  }
}

boolean sendDirectoryList(File *di) {
  if (*di) {
    di->rewindDirectory();
    client.println(F("HTTP/1.0 200 OK"));
    client.println(F("Content-Type: text/html"));
    client.println(F("Connection: close"));
    client.println();
    client.println(F("<html>"));
    client.print(F("<head><title>"));
    client.print(fname);
    client.println(F("</title></head>"));
    client.println(F("<body>"));
    client.print(F("<h1>Index of "));
    client.print(fname);
    client.println(F("</h1>"));
    client.println(F("<table cellpadding=2 cellspacing=2 border=0>"));
    File lsf;
    while ((lsf = di->openNextFile())) {
      client.println(F("<tr>"));
      client.print(F("<td>"));
      if (lsf.isDirectory())
        client.print(F("[dir]"));
      else
```

```
        client.print(lsf.size());
      client.print(F("</td>"));
      client.print(F("<td>"));
      client.print(F("<a href='"));
      client.print(fname);
      if (fname[1] != 0)
        client.print(F("/"));
      client.print(lsf.name());
      client.print(F("'>"));
      client.print(lsf.name());
      client.print(F("</a>"));
      client.print(F("</td>"));
      client.println(F("</tr>"));
      client.flush();
      lsf.close();
    }
    client.println(F("</table>"));
    client.println(F("</body>"));
    client.println(F("</html>"));
    return true;
  }
  else
    return false;
}

boolean sendFile(File *fi) {
  if (*fi) {
    client.println(F("HTTP/1.0 200 OK"));
    client.print(F("Content-Type: "));
    byte sl = strlen(fname);
    if (sl > 4) {
      if ((fname[sl-4]=='.') && (fname[sl-3]=='H') && (fname[sl-2]=='T') && (fname[sl-1]=='M'))
        client.println(F("text/html"));
      else if ((fname[sl-4]=='.') && (fname[sl-3]=='C') && (fname[sl-2]=='S') && (fname[sl-1]=='S'))
        client.println(F("text/css"));
      else if ((fname[sl-4]=='.') && (fname[sl-3]=='A') && (fname[sl-2]=='S') && (fname[sl-1]=='P'))
        client.println(F("text/html"));
      else if ((fname[sl-4]='.') && (fname[sl-3]=='P') && (fname[sl-2]=='N') && (fname[sl-1]=='G'))
        client.println(F("image/png"));
      else
        client.println(F("application/octet-stream"));
    } else if (sl > 3) {
      if ((fname[sl-3]=='.') && (fname[sl-2]=='J') && (fname[sl-1]=='S'))
        client.println(F("text/plain"));
      else
        client.println(F("application/octet-stream"));
    } else if (sl > 2) {
      if ((fname[sl-2] == '.') && ( (fname[sl-1] == 'H') || (fname[sl-1] == 'C')) )
        client.println(F("text/plain"));
      else
        client.println(F("application/octet-stream"));
    }
    else
      client.println(F("application/octet-stream"));
    client.println(F("Connection: close"));
    client.println();

    if (sl > 4) {
      if ((fname[sl-4]=='.') && (fname[sl-3]=='A') && (fname[sl-2]=='S') && (fname[sl-1]=='P')) {
        char tmpc;
        boolean found_mark1 = false;
        boolean found_mark2 = false;
        while (fi->available()) {
          tmpc = fi->read();
          if ((tmpc == '%') && (!found_mark1))
            found_mark1 = true;
          else if ((tmpc == '%') && (found_mark1))
            found_mark2 = true;
          else {
```

```
        if ((found_mark1) && (found_mark2)) {
          switch (tmpc) {
            case 'r':
              client.print(random());
              break;
          }
        }
        else if ((found_mark1) && (!found_mark2)) {
          client.print('%');
          client.write(tmpc);
        }
        else
          client.write(tmpc);
        found_mark1 = found_mark2 = false;
      }
    }
    return true;
  }
}
while (fi->available()) {
  client.write(fi->read());
}
return true;
}
else
  return false;
}

void loop() {
  client = myServer.available();

  if (client) {
    while (client.connected()) {
      if (client.available() >= 4) {
        client.read();
        client.read();
        client.read();
        client.read();
        break;
      }
    }

    if (client.connected()) {
      readFileRequest();

      if (
        (fname[0] == '/') &&
        (fname[1] == 0)
        )
      {
        File di = SD.open(fname);
        if (!sendDirectoryList(&di))
          sendError(500, F("Internal Server Error"));
        di.close();
      }
      else {
        if (SD.exists(fname)) {
          File tmp = SD.open(fname);
          if (tmp.isDirectory())
            sendDirectoryList(&tmp);
          else
            sendFile(&tmp);
          tmp.close();
        }
        else if ( (fname[0]=='/') && (fname[1]=='g') && (fname[2] == 's') && (fname[3]==0) ) {
            client.println(F("HTTP/1.0 200 OK"));
            client.println(F("Content-Type: application/json"));
            client.println(F("Cache-Control: no-cache, no-store, must-revalidate"));
            client.println(F("Pragma: no-cache"));
```

```
                    client.println(F("Expires: 0"));
                    client.println(F("Connection: close"));
                    client.println();
                    client.print(F("{\"result\":\"ok\", \"values\":["));
                    client.print(Angle1);
                    client.print(F(","));
                    client.print(Angle2);
                    client.print(F(","));
                    client.print(Angle3);
                    client.print(F(","));
                    client.print(Angle4);
                    client.print(F("]}"));
              }
          else if ((strlen(fname)>=3) && (fname[0]=='/') && (fname[1]=='s') && (fname[2]=='s') ) {
              if (strlen(fname)==3) {
                 client.println(F("HTTP/1.0 200 OK"));
                 client.println(F("Content-Type: application/json"));
                 client.println(F("Cache-Control: no-cache, no-store, must-revalidate"));
                 client.println(F("Pragma: no-cache"));
                 client.println(F("Expires: 0"));
                 client.println(F("Connection: close"));
                 client.println();
                 client.println(F("{\"result\":\"error\", \"message\":\"Invalid command\"}"));
              }
              else {
                 char buffer[4];
                 byte i;
                 for (i=4; i < strlen(fname); i++)
                   buffer[i-4] = fname[i];
                 buffer[i-4] = 0;
                 byte angle = 180 - atoi(buffer);
                 switch (fname[3]) {
                   case '0':
                     Angle1 = angle;
                     servo1.write(180 - angle);
                     break;
                   case '1':
                     Angle2 = angle;
                     servo2.write(180 - angle);
                     break;
                   case '2':
                     Angle3 = angle;
                     servo3.write(180 - angle);
                     break;
                   case '3':
                     Angle4 = angle;
                     servo4.write(180 - angle);
                     break;
                 }
                 client.print(F("{\"result\":\"ok\", \"values\":["));
                 client.print(Angle1);
                 client.print(F(","));
                 client.print(Angle2);
                 client.print(F(","));
                 client.print(Angle3);
                 client.print(F(","));
                 client.print(Angle4);
                 client.print(F("]}"));
              }
           }
           else
              sendError(404, F("File Not Found"));
        }
      }
      client.stop();
  }
}
```

Using UDP and Socket Programming

The previous projects in this book are focused on communicating with clients and servers using the hypertext transfer protocol (HTTP) over transmission control protocol (TCP) port 80. Aside from the exchange of web requests and webpage data, HTTP is also used for other purposes. In Project 3 – Building a Twitter Alarm on page 51 you can see how to exchange data that is not specifically related to webpages over HTTP. In Project 6 – Controlling Digital Outputs from the Web on page 91 you implement a small application programming interface (API) over HTTP, and return JavaScript object notation (JSON) instead of webpages.

There are many good reasons to use HTTP for applications, beyond receiving or serving webpages. One of which is that firewalls and other security measures rarely block TCP port 80, due to how commonly the web is used. In addition, most programming languages either have built-in methods for making web requests, or have well-tested libraries to perform this function.

But there are other application protocols – not many, but HTTP is not the only one – that also communicate over Internet protocols.

Socket is a fairly generic term that refers to the connection between two machines, and it implies no definition of clients or servers, or use of a specific protocol. The Arduino's Ethernet library opens sockets, and you can use these to write code that works with other application protocols.

Project 7 – Building a Local DNS Server on page 112 begins this exploration by building a server device that communicates using the domain name system (DNS) protocol to find domain names by looking up IP addresses.

Project 8 – Implementing a Custom Protocol on page 130 expands on this introduction and covers why you might want to write your own protocol, and how to do it.

In This Chapter

Project 7 – Building a Local DNS Server

As mentioned previously in this book, when computers connect over Internet protocols, they use IP addresses. They do not use domain names. To fetch a webpage, your computer needs to find out what IP address is associated with the domain name that you have entered. It asks a *nameserver*. This server tries to answer the question, but if it does not know then it asks another nameserver. If this second server does not know either then it asks a third, the third asks a fourth, and so on. This continues until either a record is found that associates the domain name with an IP address, or the request reaches a sufficiently high-level server that can say that the domain name does exist.

The way clients ask nameservers, and the way in which servers respond, is formalized in the domain name system (DNS) protocol.

The key differences between the DNS protocol and HTTP are:

- HTTP sends all of the header information and content in ASCII characters; DNS sends information in binary.

- HTTP communicates over TCP port 80; DNS communicates over UDP port 53.[1]

- UDP has a less-strict definition of clients and servers, so it is a little simpler for small devices to act as both if the circumstances demand it.

- UDP is slightly less reliable than TCP. It is not uncommon for messages to be ignored or lost.

In this project, you will write a sketch that transforms your Arduino into a nameserver. This server:

1. Connects to the network and accepts DNS requests from clients.

2. Provides IP addresses for machines on the local network, associating them with domain names that cannot exist on the Internet.

3. Asks another DNS server for information if it cannot answer a request.

Any machine on your network (or even across the Internet, if you configure your router to allow it) that uses the Arduino nameserver for its DNS will be able to use your custom domain names.

To begin, you will write a sketch that accepts DNS requests and then passes these requests to another nameserver. When the Arduino receives a response, it passes the message back to the machine that originally made the request.

---

1. User datagram protocol (UDP) is explained in the next section.

For these early stages, you only need a very small amount of knowledge about the structure of DNS messages. The remaining information is introduced to you in later sections, when the sketch is modified to examine the requests from clients and send back its own responses. However, the protocol specification is also in the appendix – DNS – Implementation and Specification on page 182.

Waiting for UDP Connections

UDP is not TCP, and UDP port 53 is not the same as TCP port 53. In preparation for writing the server, you should refer to Project 4 – Setting up a Basic Web Server on page 65 to learn how to use a static IP address and port forwarding. In particular, if you want to allow connections from outside of your local area network then you will need to configure your router so that it forwards UDP port 53 to the Arduino.

The Ethernet library includes a separate class for communicating over UDP, and requires an additional include directive at the top of the sketch.

```
#include <EthernetUDP.h>
```

Then declare an instance of the *EthernetUDP* class:

```
EthernetUDP udp;
```

To start the library code and monitor a port, use the *begin()* method of the *EthernetUDP* class. This accepts one argument – the UDP port number to communicate on.

The basic sketch below connects to the network and declares an instance of the *EthernetUDP* class. You will build on this sketch as you progress through the project, until you have a working DNS server.

```
#include <SPI.h>
#include <Ethernet.h>
#include <EthernetUDP.h>

const byte mac[] = { 0x00, 0xC3, 0xA2, 0xE6, 0x3D, 0x54 };
byte ip[] = { 192, 168, 0, 99 };
IPAddress dns_server(8,8,8,8);
byte pbuf[512];
EthernetUDP udp;

void setup() {
  //D53 on the Arduino Mega must be an output.
  pinMode(53, OUTPUT);

  Serial.begin(9600);
  while (!Serial);

  Serial.println("Establishing network connection...");
```

```
    Ethernet.begin((uint8_t*)mac, (uint8_t*)ip);

    Serial.print("IP Address: ");
    Serial.println(Ethernet.localIP());

    Serial.print("Opening UDP port... ");
    if (udp.begin(53) == 1)
      Serial.println("OK!");
    else
      Serial.println("FAILED!");
}

void loop() {
}
```

The global variable declarations will be familiar to you if you have completed the other projects in this book. The sketch declares an additional IP address, *dns_server*, which specifies the IP address of a public DNS server to which the Arduino can relay DNS requests. 8.8.8.8 is the address of Google's public DNS server.

The next step is to write code into the sketch's *loop()* function so that it waits until another machine sends a UDP message to the Arduino.

The method *parsePacket()* checks for the presence of a suitable message, and must be called before data can be read into a buffer array using *read()*.

```
void loop() {
  int psize = udp.parsePacket();
  if (psize > 0) {
    udp.read(pbuf, sizeof(pbuf));
  }
}
```

read() accepts two arguments. The first is a pointer to an area of memory in which to store the bytes read from the message. The second is the maximum number of bytes to read – the value returned by the function indicates how many bytes are actually read.

Now that the DNS message is stored in a buffer array, you can process the request.

In the next section, you will modify the sketch's *loop()* function so that when the Arduino receives a DNS request, it forwards this request to another DNS server. When the Arduino receives a response to that request, it will forward the response to the original client.

Communicating over UDP Sockets and Sending a DNS Request

At this stage, you only need to know about the first four bytes of the DNS message format.

Byte	Type	Name	Description
0–1	Unsigned integer	Transaction ID	A 16-bit reference number that is used by the client to link DNS responses to the original request.
2–3	Unsigned integer	Flags	The most significant bit specifies whether the DNS message is a request (0) or a response (1).

Create a new function in the sketch:

```
int echo_DNS_Lookup(int sz) {
  int tid = random(0xFFFF);
  pbuf[0] = (byte)(tid >> 8);
  pbuf[1] = (byte)(tid & 0x0000FFFF);
  udp.beginPacket(dns_server, 53);
  udp.write(pbuf, sz);
  udp.endPacket();
  return 0;
}
```

The parameter *sz* specifies how long the DNS request is, since the allocated size of the buffer *pbuf* is often longer than the client's request.

When the Arduino forwards the request to another DNS server, it needs to be able to identify when it receives the response message that matches the request. To do this, the code above changes the transaction ID of the DNS request to a random 16-bit number.

> **Tip: When conforming to the DNS protocol, the most-significant byte of a 16-bit integer should be sent first. The Arduino is little-endian and so the two lines that set the bytes in pbuf ensure that the two halves of the random transaction ID are divided correctly.**

After a call to the *EthernetUDP* class method *begin()*, you can begin writing DNS messages to the UDP port. You do not need to open a connection to a server. Instead, the intended recipient of the message is specified in the message itself.

This code uses three methods of the *EthernetUDP* class:

Method	Description
beginPacket()	Begins a UDP message, sending the IP address of the intended recipient, and the port number.
write()	Writes a series of bytes to the UDP message. The second parameter is the number of bytes to send.
endPacket()	Finish sending the UDP message.

After sending the message, the *echo_DNS_Lookup()* function should wait for a DNS response message that has the same transaction ID. Add the following code before the line *return 0;* in *echo_DNS_Lookup()*:

```
long timeout = millis() + 1000;
while (true) {
  int result = udp.parsePacket();
  if (result > 0) {
    udp.read(pbuf, sizeof(pbuf));
    if ( (pbuf[2] & 0x80) == 0x80) &&
        (pbuf[0] == ((byte)(tid >> 8))) &&
        (pbuf[1] == ((byte)(tid & 0x0000FFFF)))
      )
      return result;
    if (millis() > timeout)
      return 0;
  }
  delay(10);
}
```

To ensure that only DNS response messages are processed, this code checks whether bit 15 of the message flags is set. If it is, the message is a DNS response and the loop can end.

parsePacket() does not return until a UDP message is found, and so the timeout that is implemented here is a little crude, but functional.

When the *echo_DNS_Lookup()* function exits, it either returns *0* to tell the calling function that no response was received from the DNS server, or it returns the length of the response. The response is in the buffer array *pbuf* – this overwrites the client's original DNS request, but at this point that is no longer needed.

Returning a DNS Record

To modify the sketch's *loop()* function to call *echo_DNS_Lookup()* when the Arduino receives a DNS request:

1. Check whether bit 15 of the DNS flags is clear – a request.

2. If it is, store the client's IP address, remote port number, and the request's transaction ID.

3. Call *echo_DNS_Lookup()* to send the DNS request to another DNS server.

4. If the result is not *0*, change the transaction ID back to match the client's original request.

5. Send the DNS response to the client using the stored IP and remote port.

You can do this by adding the following code to sketch's *loop()* function, after the call to *udp.read()*:

```
if ((pbuf[2] & 0x80) == 0) {
  IPAddress client = udp.remoteIP();
  unsigned int clientPort = udp.remotePort();
  int tid = (pbuf[0] << 8) | pbuf[1];
  int res = echo_DNS_Lookup(psize);
  if (res > 0) {
    pbuf[0] = (byte)(tid >> 8);
    pbuf[1] = (byte)(tid & 0x0000FFFF);
    udp.beginPacket(client, clientPort);
    udp.write(pbuf, res);
    udp.endPacket();
  }
}
```

At this point, the entire sketch should look something like:

```
#include <SPI.h>
#include <Ethernet.h>
#include <EthernetUDP.h>

const byte mac[] = { 0x00, 0xC3, 0xA2, 0xE6, 0x3D, 0x54 };
byte ip[] = { 192, 168, 0, 99 };
IPAddress dns_server(8,8,8,8);
byte pbuf[512];
EthernetUDP udp;

void setup() {
  //D53 on the Arduino Mega must be an output.
  pinMode(53, OUTPUT);

  Serial.begin(9600);
  while (!Serial);

  Serial.println("Establishing network connection...");

  Ethernet.begin((uint8_t*)mac, (uint8_t*)ip);

  Serial.print("IP Address: ");
  Serial.println(Ethernet.localIP());
```

```
  Serial.print("Opening UDP port... ");
  if (udp.begin(53) == 1)
    Serial.println("OK!");
  else
    Serial.println("FAILED!");
}

int echo_DNS_Lookup(int sz) {
  int tid = random(0xFFFF);
  pbuf[0] = (byte)(tid >> 8);
  pbuf[1] = (byte)(tid & 0x0000FFFF);
  udp.beginPacket(dns_server, 53);
  udp.write(pbuf, sz);
  udp.endPacket();

  long timeout = millis() + 1000;
  while (true) {
    int result = udp.parsePacket();
    if (result > 0) {
      udp.read(pbuf, sizeof(pbuf));
      if ( ((pbuf[2] & 0x80) == 0x80) &&
           (pbuf[0] == ((byte)(tid >> 8))) &&
           (pbuf[1] == ((byte)(tid & 0x0000FFFF)))
         )
        return result;
      if (millis() > timeout)
        return 0;
    }
    delay(10);
  }
  return 0;
}

void loop() {
  int psize = udp.parsePacket();
  if (psize > 0) {
    udp.read(pbuf, sizeof(pbuf));

    if ((pbuf[2] & 0x80) == 0) {
      IPAddress client = udp.remoteIP();
      unsigned int clientPort = udp.remotePort();
      int tid = (pbuf[0] << 8) | pbuf[1];

      int res = echo_DNS_Lookup(psize);
      if (res > 0) {
        pbuf[0] = (byte)(tid >> 8);
        pbuf[1] = (byte)(tid & 0x0000FFFF);
        udp.beginPacket(client, clientPort);
        udp.write(pbuf, res);
        udp.endPacket();
      }
    }
  }
  delay(10);
}
```

Testing the Device

The nslookup tool can help you test the DNS server, without setting the Arduino as your computer's primary DNS system.

On Windows 8/7/Vista/XP:

1. Press the **Windows logo key + R**.

2. Type *cmd*, then press **Enter**.

3. Type *nslookup www.connectingarduino.com 192.168.0.99* and press **Enter**. Change the IP address if your Arduino connects to your network using a different IP address.

Nslookup makes DNS requests for the specified domain, and if an additional server name or IP is included, it will use the specified nameserver. You should see a response similar to Figure 17.

Figure 17. A successful DNS lookup on Windows

On Mac OS X:

1. On the dock, click **Finder**.

2. On the sidebar, click **Applications**.

3. Click **Utilities**, then double-click **Terminal**.

4. Type n*slookup www.connectingarduino.com -server 192.168.0.99* and press **Enter**. Change the IP address if your Arduino connects to your network using a different IP address.

If the Arduino sketch is functioning correctly, then you should see very little difference between running nslookup without specifying the nameserver and when you run it using the IP address of the Arduino.

In the next section, you will see the structure of DNS requests and responses. One of the pieces of information that a DNS server adds to its response is the time-to-live value (TTL). This indicates how long DNS clients are allowed to cache the record before they should request a new copy from the nameserver. If you find that nslookup is caching the information, preventing you from testing effectively while you work on this project, you can clear the DNS cache.

On Windows 8/7/Vista/XP:

1. Press the **Windows logo key + R**.
2. Type *cmd*, then press **Enter**.
3. Type *ipconfig /flushdns* and then press **Enter**.

On Mac OS X (Tiger and earlier):

1. On the dock, click **Finder**.
2. On the sidebar, click **Applications**.
3. Click **Utilities**, then double-click **Terminal**.
4. Type *lookupd -flushcache* and then press **Enter**.

On Mac OS X (Leopard and later):

1. On the dock, click **Finder**.
2. On the sidebar, click **Applications**.
3. Click **Utilities**, then double-click **Terminal**.
4. Type *dscacheutil -flushcache* and then press **Enter**.

Implementing Custom Domain Names and Routing

The aim of this project is not simply to act as proxy server. In this section, you will modify the sketch so that it responds to DNS requests itself. To do this, you need to understand the DNS requests.

This is the structure of a header in a DNS message:

Byte	Type	Name	Description
0–1	Unsigned integer	Transaction ID	A 16-bit reference number that is used by the client to link DNS responses to the original request

Byte	Type	Name	Description
2–3	Unsigned integer	Flags	See below.
4–5	Unsigned integer	Questions	The number of domains to be resolved to IP addresses. Usually *1*.
6–7	Unsigned integer	Answer RRs	Typically *0* for requests.
8–9	Unsigned integer	Authority RRs	Typically *0* for requests.
10–11	Unsigned integer	Additional RRs	Typically *0* for requests.
12…		Queries	See below.

The *flags* field packs 10 pieces of information in a 16-bit structure. Only five items are needed for DNS requests – the other bits are clear.

Bit	Name	Description
15	QR	Clear if the DNS message is request, set if the DNS message is a response.
14–11	Opcode	A standard DNS request is *0*. The value *1* indicates an inverse lookup, and *2* indicates a DNS status request.
9	TC	Truncation. Specifies that the DNS message was too long for a single UDP message and is split across multiple. This project assumes requests and responses are not truncated (*0*).
8	RD	Recursion desired. *1* specifies that the DNS server should contact other DNS servers if it needs to, and is the most-common value.
6–4	Z	Reserved. Should be *0*.

After the header structure, which is always 12 bytes, the client sends a number of query structures, the exact number of which is set in the header field *Questions*.

Each query is of variable length, due to the way the domain names are included, but they are made up of three sections: QNAME (the domain name), QTYPE (the type of DNS record being requested – 0x0001 for an A record that translates domain names to IP addresses), and QCLASS (0x0001 for Internet queries).

The domain name is divided into pieces, corresponding to where the period is placed when typed into a web browser. Before each piece, the client sends a byte that specifies how many characters this part of the domain name consists of. For example:

```
0x03 0x77 0x77 0x77
Length: 3
Characters: www
```

These pieces are sent one after the other until the number that specifies the length of a part is zero. The following sequence represents the domain name *www.arduino.cc*:

```
0x03 0x77 0x77 0x77 0x06 0x61 0x72 0x64 0x75 0x69 0x6e 0x6f 0x02 0x63 0x63 0x00
```

Add this function to your Arduino sketch:

```
String getDomainName() {
  String result;
  int i=12;
  while (pbuf[i] > 0) {
    if (result.compareTo("") != 0)
      result.concat('.');
    for (int x=1; x <= pbuf[i]; x++);
      result.concat((char)pbuf[i+x]);
    i = i + pbuf[i] + 1;
  }
  return result;
}
```

The first QNAME of the query will always be at position 12 (starting at the 13th byte) in the buffer array, so this function reads the QNAME and returns a *String* object that contains the domain name in a usual period-separated way.

At the top of the sketch, define a few domain names:

```
const int NUM_HOSTS = 3;
String hosts[] = {"dns.arduino", "my.arduino", "web.arduino"};
String ips[] = { IPAddress(192,168,0,99), IPAddress(192,168,0,10), IPAddress(192,168,0,11) };
```

These domain names cannot exist on the Internet since *.arduino* is not a valid top-level domain.

Now, add this function to your sketch:

```
int isLocalDomain(String dn) {
  for (int i=0; i < NUM_HOSTS; i++) {
    if (dn.equalsIgnoreCase(hosts[i]))
      return i;
  }
  return -1;
}
```

isLocalDomain() returns *-1* if the domain name passed as the string argument *dn* is not in the list of custom domain names. If it is in the list, *isLocalDomain()* returns the position of the domain in the array.

You need to change your sketch's *loop()* function so that it:

1. Checks that the request's opcode (bits 11–14) is *0*.

2. Checks if the domain is in the list of local domains.

3. Sends the request to another DNS server, using *echo_DNS_Lookup()*, if the domain is not in the list of local domains.

To do this, change *loop()* so that it looks like:

```
void loop() {
  int psize = udp.parsePacket();
  if (psize > 0) {
    udp.read(pbuf, sizeof(pbuf));

    if ((pbuf[2] & 0x80) == 0) {
      if ((pbuf[2] & 0x78) == 0) {
        IPAddress client = udp.remoteIP();
        unsigned int clientPort = udp.remotePort();
        int tid = (pbuf[0] << 8) | pbuf[1];

        int host_idx = isLocalDomain(getDomainName());
        if (host_idx == -1) {
          int res = echo_DNS_Lookup(psize);
          if (res > 0) {
            pbuf[0] = (byte)(tid >> 8);
            pbuf[1] = (byte)(tid & 0x0000FFFF);
            udp.beginPacket(client, clientPort);
            udp.write(pbuf, res);
            udp.endPacket();
          }
        } else {
          // Send a DNS response to the client here.
        }
      }
    }
  }
  delay(10);
}
```

Test the sketch as described in Testing the Device on page 118. When querying local domains, nslookup should timeout without receiving any response (see Figure 18). For all other domains, you should receive a DNS response.

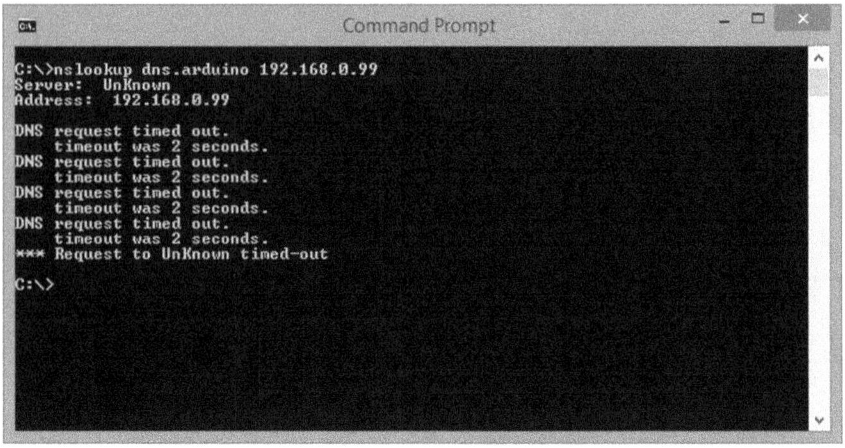

Figure 18. No response from the Arduino server

The final step in this project is to replace the comment *// Send a DNS response to the client here*, with code that sends a DNS response message to the client.

In Implementing Custom Domain Names and Routing on page 120 you can see the structure of a DNS message, in particular, the structure of the header. Response messages follow the same format as request messages – except that after the queries section, the server sends up to three sets of answers: answers, authoritative nameservers, and additional records. Each set can contain multiple records. However, in this project you will only send one and so the field *Answer RRs* (bytes 6–7) in the message header will be an unsigned integer with the value *1*.

Answer records follow a standard format. Since the domain name is variable in length, the byte numbers in the table below are shown assuming that +1 is the byte following the domain name.

Byte	Type	Name	Description
0...		NAME	The domain name to which this record pertains.
+1	Unsigned integer	TYPE	Specifies the type of record returned. This project only sends A records (type 0x0001). See 3.2.2 TYPE values on page 191.
+3	Unsigned integer	CLASS	The class of data. 0x0001 for Internet records.
+5	32-bit unsigned integer	TTL	Specifies how long (in seconds) the browser is permitted to cache this DNS record for.

Byte	Type	Name	Description
+9	Unsigned integer	RDLENGTH	The number of bytes of data. This is 4 when returning IP addresses in A name records.
+11		RDATA	The IP address as four bytes.

The code below:

1. Sends the transaction ID.

2. Sends 10 bytes that are the same for every DNS response it returns – including a standard, no-error flags field.

3. Copies the client's original queries block to the DNS response.

4. Sends the answer record – including the IP address of the domain being queried, and a value indicating that the record can be cached for three minutes (180 seconds).

When sending the answer record, the code copies the domain name from the client's request for use in the NAME field.

Replace the line *// Send a DNS response to the client here* with the following code. The full sketch is shown in Source Code on page 127.

```
udp.beginPacket(client, clientPort);

// send transaction id
udp.write((byte)(tid >> 8));
udp.write((byte)(tid & 0x0000FFFF));

// send standard header info
byte std_info[] = {0x81, 0x80, 0, 1, 0, 1, 0, 0, 0, 0};
udp.write(std_info, 10);

// copy the queries block from the request
int ix = 12;
while (pbuf[ix] != 0)
  udp.write((byte)pbuf[ix++]);
for (int x=ix; x < ix+5; x++)
  udp.write((byte)pbuf[x]);

// copy the domain name from the request
ix = 12;
while (pbuf[ix] != 0)
  udp.write((byte)pbuf[ix++]);

// send a second set of standard values
byte std_info2[] = {0, 0, 1, 0, 1, 0, 0, 0, 180, 0, 4};
udp.write(std_info2, 11);
udp.write((byte)ips[host_idx][0]);
udp.write((byte)ips[host_idx][1]);
udp.write((byte)ips[host_idx][2]);
udp.write((byte)ips[host_idx][3]);

udp.endPacket();
```

The Arduino now returns answer records that DNS clients can understand.

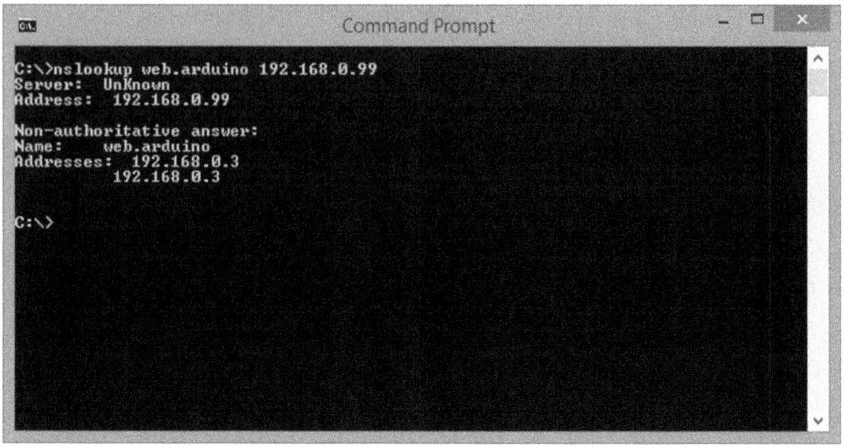

Figure 19. Resolution of a made-up domain to its IP address

Embellishing the Project

UDP is typically a little less reliable than TCP, but this can be an advantage for Arduino-based DNS servers because clients are forced to deal with the possibility that they will not receive a response.

In this project, while the sketch is waiting for a response from another DNS server, it does not process any incoming requests from clients. However, clients usually only wait a second or two before sending another request if their previous attempt is unanswered. This gives the echo_DNS_Lookup() function time to exit and return to the main loop in loop().

Regardless of its lack of support for simultaneous connections, you now have a working DNS server that can be used by any of your Internet-enabled computers and devices. Although the project is complete, there are two enhancements you can make that would greatly improve its usefulness.

Inverse lookups (often called reverse lookups) perform a different kind of DNS operation. Instead of the client sending a domain name and expecting an IP address, the client sends an IP address and hopes that the server responds with a record that includes the domain name. This type of request is marked with an opcode of 1 in the DNS message header, and by using a domain name that includes the IP address followed by IN-ADDR.ARPA. For the local domains processed in this project, it is relatively straightforward to extend the sketch to support to inverse lookups. For more information about inverse lookups and the IN-ADDR.ARPA domain, see the entry in Appendix B, section 3.5 IN-ADDR.ARPA domain on page 198.

Finally, you might consider modifying the sketch's *setup()* function so that it reads a list of domain names and IP address from the SD card. This would save you from having to reprogram the Arduino if you add more devices and domain names to the system.

Source Code

The full source code for this sketch is shown below. If you want to define your own configuration[1] then the declarations to change are:

Name	Description
mac	The device's unique MAC address as an array of bytes.
ip	The IP address that the sketch uses when connecting to the network.
dns_server	Defines the IP address of a nameserver to use when the Arduino cannot answer the DNS request itself.
NUM_HOSTS	The number of entries in the *hosts* and *ips* arrays.
hosts	An array of *String* objects that defines the domain names that the Arduino will interpret itself.
ips	An array of *IPAddress* objects. These are the IP addresses that match the entries in the hosts array.

```
#include <SPI.h>
#include <Ethernet.h>
#include <EthernetUDP.h>

const byte mac[] = { 0x00, 0xC3, 0xA2, 0xE6, 0x3D, 0x54 };
byte ip[] = { 192, 168, 0, 99 };
IPAddress dns_server(8,8,8,8);
byte pbuf[512];
EthernetUDP udp;

const int NUM_HOSTS = 3;
String hosts[] = {"dns.arduino", "my.arduino", "web.arduino"};
IPAddress ips[] = { IPAddress(192,168,0,99), IPAddress(192,168,0,2), IPAddress(192,168,0,3) };

void setup() {
  //D53 on the Arduino Mega must be an output.
  pinMode(53, OUTPUT);

  Serial.begin(9600);
  while (!Serial);

  Serial.println("Establishing network connection...");
```

---

1. By now, you should be very familiar with setting the MAC and IP address for how the Arduino connects to the network.

```
  Ethernet.begin((uint8_t*)mac, (uint8_t*)ip);

  Serial.print("IP Address: ");
  Serial.println(Ethernet.localIP());

  Serial.print("Opening UDP port... ");
  if (udp.begin(53) == 1)
    Serial.println("OK!");
  else
    Serial.println("FAILED!");
}

int echo_DNS_Lookup(int sz) {
  int tid = random(0xFFFF);

  //change transaction id
  pbuf[0] = (byte)(tid >> 8);
  pbuf[1] = (byte)(tid & 0x0000FFFF);
  udp.beginPacket(dns_server, 53);
  udp.write(pbuf, sz);
  udp.endPacket();

  long timeout = millis() + 1000;
  while (true) {
    int result = udp.parsePacket();
    if (result > 0) {
      udp.read(pbuf, sizeof(pbuf));
      if (
          ((pbuf[2] & 0x80) == 0x80) &&
          (pbuf[0] == ((byte)(tid >> 8))) &&
          (pbuf[1] == ((byte)(tid & 0x0000FFFF)))
         )
        return result;
      if (millis() > timeout)
        return 0;
    }
    delay(10);
  }
  return 0;
}

void loop() {
  int psize = udp.parsePacket();
  if (psize > 0) {
    udp.read(pbuf, sizeof(pbuf));

    if ((pbuf[2] & 0x80) == 0) {
      if ((pbuf[2] & 0x78) == 0) {
        IPAddress client = udp.remoteIP();
        unsigned int clientPort = udp.remotePort();
        int tid = (pbuf[0] << 8) | pbuf[1];

        int host_idx = isLocalDomain(getDomainName());
        if (host_idx == -1) {
          int res = echo_DNS_Lookup(psize);
          if (res > 0) {
            pbuf[0] = (byte)(tid >> 8);
            pbuf[1] = (byte)(tid & 0x0000FFFF);
            udp.beginPacket(client, clientPort);
            udp.write(pbuf, res);
            udp.endPacket();
          }
        } else {
          udp.beginPacket(client, clientPort);
          udp.write((byte)(tid >> 8));
          udp.write((byte)(tid & 0x0000FFFF));

          byte std_info[] = {0x81, 0x80, 0, 1, 0, 1, 0, 0, 0, 0};
          udp.write(std_info, 10);
```

```
          int ix = 12;
          while (pbuf[ix] != 0)
            udp.write((byte)pbuf[ix++]);
          for (int x=ix; x < ix+5; x++)
            udp.write((byte)pbuf[x]);

          ix = 12;
          while (pbuf[ix] != 0)
            udp.write((byte)pbuf[ix++]);
          byte std_info2[] = {0, 0, 1, 0, 1, 0, 0, 0, 180, 0, 4};
          udp.write(std_info2, 11);
          udp.write((byte)ips[host_idx][0]);
          udp.write((byte)ips[host_idx][1]);
          udp.write((byte)ips[host_idx][2]);
          udp.write((byte)ips[host_idx][3]);

          udp.endPacket();
        }
      }
    }
  }
  delay(10);
}

String getDomainName() {
  String result;
  int i = 12;
  while (pbuf[i] > 0) {
    if (result.compareTo("") != 0)
      result.concat('.');
    for (int x=1; x <= pbuf[i]; x++)
      result.concat((char)pbuf[i+x]);
    i = i + pbuf[i] + 1;
  }
  return result;
}

int isLocalDomain(String dn) {
  for (int i=0; i < NUM_HOSTS; i++) {
    if (dn.equalsIgnoreCase(hosts[i]))
      return i;
  }
  return -1;
}
```

Project 8 – Implementing a Custom Protocol

The transmission control protocol (TCP) encompasses everything that is needed to ensure that messages are delivered to the intended recipient, and all of this is handled by the Arduino's Ethernet library and the Ethernet Shield. Much like working with a hardware serial port, you can open a TCP socket on a specified port and then write strings and binary data. If the connection is held open, the recipient can respond in the same way.

However, if the messages need to be structured and interpreted by both parties, then you need an application protocol – an agreement between the client and server (or simply any two or more machines) that defines how data should be represented when it is sent.

If you have read the previous projects in this book, then you have already seen how the hypertext transfer protocol (HTTP) and domain name system (DNS) protocols work. You have also seen that HTTP is extremely flexible.

But there are occasions when you can consider writing your own protocol. This can only be done when you have control of both ends of the message – you must be able to define the functionality of the client as well as the server – but doing so has a few benefits for Arduino programmers:

- Overhead can be reduced. For example, HTTP is not optimized for fast transmission – being strings, the HTTP header fields can be almost as long as the data being exchanged. The relatively slow processing speed and transmission speed of the Arduino and the Ethernet Shield mean that trimming down the amount of additional information will significantly increase the speed at which the actual data is sent.

- Ease of processing. With the Arduino's limited processing power and memory, considerable gains in performance can be found by designing a protocol that is easy for the Arduino to process.

- Exclusivity. Sometimes it is beneficial to make it so that other clients cannot communicate using your protocol without having first been specifically adapted for that purpose.[1]

- Unique requirements. A project may require that data is exchanged in a way that no existing protocol can adequately support.

This is more of a guide than a project, looking at some of the issues that you should consider when designing and implementing a custom protocol. It also includes more examples of how to send and receive data over TCP using the Ethernet library, and a short section with an example of how to write client software for your PC.

---

1. Be aware that protocols can be, and often are, reverse engineered. Obscurity is not security.

In addition to the Arduino integrated development environment, you will need to download and install Processing to work with the client code.[1]

Defining a Protocol

The first step in designing a protocol is to define what data needs to be exchanged between the client and the server (or between machines, if you are building a system in which the traditional roles are of little use – such as a peer-to-peer system).

Over the next few sections, you will write an implementation of a protocol that gives basic remote control over the Arduino's digital pins and analog inputs to a program running on your PC. This protocol is called *remote control of [Ar]duino inputs and outputs* (RCDIO).[2]

The client software will send commands to the Arduino, and in this example the commands are based on the equivalent functions that you use when programming an Arduino sketch.

Command	Arduino Equivalent	Description
0x01	–	Status. Returns information about the server – such as the type of Arduino it is connected to. Accepts no arguments.
0x02	pinMode()	Configure the specified pin as an *input* or *output*. Requires a pin number and the mode as arguments.
0x03	digitalWrite()	Sets the state of the specified pin as *high* or *low* depending on the second value passed in. Accepts two arguments: the pin number, and a value.
0x04	digitalRead()	Returns the state of the specified pin – whether it is *high* or *low*. Accepts one argument (pin number) and returns the state.
0x05	analogWrite()	Uses pulse width modulation (PWM) to send an analog value through the specified digital pin. Accepts two arguments: the pin number, and a value.
0x06	analogRead()	Reads the value from the specified analog input pin. Accepts one argument (pin number) and returns an integer 0–1023.

---

1. Processing is a programming language and development environment that is often used by Arduino programmers. You can download it, for free, at www.processing.org
2. RCDIO is not an existing protocol, it was created for this book.

The commands 0x02, 0x03, and 0x05 do not need to send back a response, other than to say that the command was successful. The commands 0x04 and 0x06 need to return an integer value. Command 0x01 will return a string.

From looking at the commands this protocol supports, you can see that it also needs to define how pin numbers, *high* and *low*, *input* and *output*, strings, and integer values are represented.

Type	Representation/Implementation Notes
Pin number	Sent as a single byte value 0–255.
High/low	Sent as a single byte. A value of *0* corresponds to *low*, *1* corresponds to *high*.
Input/output	Sent as a single byte. A value of *0* corresponds to *input*, *1* corresponds to *output*.
String	As the command 0x01 is the only command that uses strings (and it only returns one as its result), they are sent as a sequence of bytes; are not null-terminated; and give no indication of their length.
Integer	Sent as a 16-bit unsigned integer – two bytes with the least-significant byte first.

Now you need to define the format of messages. In this protocol, messages will be sent from the client to the server, and from the server to the client in the following format:

Byte	Name	Description
0–2	Reserved	Three ASCII characters with the values 'R', 'C', and 'D'.
3	Opcode	A single byte that states whether the message is a response from the server (*0*) or the number of one of the six commands defined earlier.
4	Length	A single byte that specifies how many bytes of data are included after the message header.
5...	Parameters	Each command argument will be sent one after the other as a byte sequence of up to 255 bytes. It is up to the application to understand how the sequence is broken up into individual arguments (based on the opcode).

This protocol will run on TCP port 80. You should avoid using ports that are reserved by other protocols unless you are sure that no other software on your computer is listening for connections on that port number. However, as port 80 is rarely blocked by firewalls and other security measures it is being used here.

The client opens the connection and sends a command request to the Arduino. If the server can process the request successfully, it sends back a response message (opcode 0) with either no data or the return value of the command. Finally, the server closes the connection.

In the event of an error, the server closes the connection without returning a response. If the client implements a timeout then it can try sending the request again after a few seconds.

Building the Server

To begin, start a new sketch that connects to the network and waits for incoming connections from clients on TCP port 80. For example:

```
#include <SPI.h>
#include <Ethernet.h>

const byte mac[] = { 0x00, 0xC3, 0xA2, 0xE6, 0x3D, 0x57 };
byte ip[] = { 192, 168, 0, 99 };

EthernetServer myServer(80);

void setup() {
  //D53 on the Arduino Mega must be an output.
  pinMode(53, OUTPUT);

  Serial.begin(9600);
  while (!Serial);

  Serial.println("Establishing network connection...");
  Ethernet.begin((uint8_t*)mac, ip);
  Serial.print("IP Address: ");
  Serial.println(Ethernet.localIP());

  myServer.begin();
}

void loop() {
  EthernetClient client = myServer.available();
  if (client) {
    client.stop();
  }
}
```

The header information in RCDIO messages consists of a total of five bytes, followed by any arguments. The initial five bytes of the header can be stored in a struct – this simplifies reading the header information sent by the client, allowing it to be done by one call to the *EthernetClient* class method *read()*.

Add the command header structure to the sketch by including the following lines at the top of the sketch:

```
struct cmd {
  char reserved[3];
  byte opcode;
  byte length;
};
```

Now modify the sketch's *loop()* function. On the first line of *loop()*, before an instance of *EthernetClient* is created, add this line to declare memory space for an instance of the *cmd* structure.

```
cmd myCmd;
```

Before the line *client.stop();* add

```
while (client.connected()) {
  if (client.available() >= 5) {
    client.read((uint8_t*)&myCmd, sizeof(myCmd));
    break;
  }
}
```

The RCDIO protocol states that the first three characters of any message should be 'R', 'C' and 'D'. Check that this is received by adding this if statement before *client.stop();*

```
if (
    (myCmd.reserved[0] == 'R') &&
    (myCmd.reserved[1] == 'C') &&
    (myCmd.reserved[2] == 'D')
  ) {
  // process command requests here
}
```

The command that the client is asking to run is a byte value in the header structure – *myCmd.opcode* – and can be processed using a switch statement. In the final sketch (Source Code – Arduino Sketch on page 138) the implementations of all of the commands are very similar.

To add support for the command 0x03 (the equivalent of the Arduino's *digitalWrite()* function), replace the comment *// process command requests here* with this code block:

```
switch (myCmd.opcode) {
  case 3:
    while (client.available() < 2);
    digitalWrite(client.read(), client.read());
    sendResponse(&client);
    break;
}
```

Command 0x03 accepts two bytes from the client, and so the code above waits until at least two bytes are available in the Ethernet Shield's buffer. These are then read and passed to *digitalWrite()*.

Finally, the code above sends back a default response message with no data. The full source code contains three forms of the function *sendResponse()* for use depending on what type of data (if any) is to be sent back to the client. The version used by command 0x03 looks like this:

```
void sendResponse(EthernetClient* client) {
  cmd response;
  response.reserved[0] = 'R';
  response.reserved[1] = 'C';
  response.reserved[2] = 'D';
  response.opcode = 0;
  response.length = 0;
  client->write((uint8_t*)&response, sizeof(response));
}
```

The function accepts a pointer to the instance of the *EthernetClient* class, and then declares a instance of the *cmd* structure. After populating the response variable, the entire header is sent to the client using a single call to the *EthernetClient* class method *write()*.

In the versions of *sendResponse()* that do return data, the length (in bytes) of the data is included in the header (*response.length*) and then the function sends the data immediately after writing the header.

Once the response is sent, the function returns and the next instruction to be executed is the call to *client.stop()*, to close the connection.

In an actual project, you should implement far more error checking than the example code does. Since RCDIO does not mandate checking for errors, it is omitted from this project so that you can see how much simpler it is for the Arduino to work with this protocol than HTTP.

Building the Client

If you choose to use a custom protocol for communication with your Arduino sketch then you will also have to write the client software, since no other software will support it. In this example, you will build a simple sketch in Processing that defines six functions (one for each of the six RCDIO commands).

Start a new sketch in Processing.

To make requests to servers over TCP, Processing sketches use the class *Client* from the *processing.net* namespace. Add an import directive to the top of the sketch:

```
import processing.net.*;
```

Because Processing does not support structs, if you want to use a similar way of handling the header information as you have in the Arduino sketch, the *cmd* structure can be implemented as a class.

```
public class cmd {
  char[] reserved = new char[3];
  byte opcode;
  byte oplength;

  public cmd() {
  }

  public cmd(byte[] fromByteArray) {
    reserved[0] = (char)fromByteArray[0];
    reserved[1] = (char)fromByteArray[1];
    reserved[2] = (char)fromByteArray[2];
    opcode = fromByteArray[3];
    oplength = fromByteArray[4];
  }

  public byte[] toByteArray() {
    byte[] result = new byte[5];
    result[0] = (byte)reserved[0];
    result[1] = (byte)reserved[1];
    result[2] = (byte)reserved[2];
    result[3] = opcode;
    result[4] = oplength;
    return result;
  }
}
```

The additional constructor, which accepts an array of bytes, and the method *toByteArray()* are included to make converting a buffer array to a *cmd* object a little easier.

Now define a few constants and declare the IP address of the Arduino server:

```
final int HIGH = 1;
final int LOW = 0;
final int OUTPUT = 1;
final int INPUT = 0;
final String serverIP = "192.168.0.99";
```

Each of the six command functions follow the same process:

1. Create an instance of the *cmd* class, and set its properties to contain the command's opcode and the length of any data (oplength) that will be sent with the request.

2. Open a connection to the server, by creating a new instance of the *Client* class.

3. Write the header structure and any data to the client, using the *Client* class method *write()*.

4. Wait until the server has sent at least five bytes of its response.

5. Convert the bytes from the response into an instance of the *cmd* class.

6. Check that the opcode of the response is *0*.

7. If the command function in the Processing sketch requires any data from the response, wait until the appropriate number of bytes has been sent.

8. Read the data, and interpret it accordingly.

9. Close the *Client* instance using the method *stop()*.

10. Exit the function, returning a value if necessary.

For example, add this function to the sketch to support sending command 0x03 requests:

```
void digitalWrite(int pin, int mode) {
  cmd myCmd = new cmd();
  myCmd.reserved[0] = 'R';
  myCmd.reserved[1] = 'C';
  myCmd.reserved[2] = 'D';
  myCmd.opcode = 3;
  myCmd.oplength = 2;

  Client c = new Client(this, serverIP, 80);
  c.write(myCmd.toByteArray());

  // send the pin number argument
  c.write((byte)pin);

  // send the mode
  if (mode == HIGH)
    c.write((byte)1);
  else
    c.write((byte)0);

  // wait for a response
  while (c.available() < 5);
  byte[] hdr = new byte[5];
  c.readBytes(hdr);
  cmd response = new cmd(hdr);

  c.stop();
}
```

As *digitalWrite()* does not require a response, the code in this example neglects to the check the opcode in the response header, and does not check for any data.

> **Tip: If you want to test the sketch at this point, you will need to add a call to *pinMode()* in the Arduino sketch to set the pins you are testing to output. Once the command 0x02 is supported by both the client and server, you will be able to change the mode of the pins on the Arduino from the client.**

If you are unsure how the other command functions can be implemented, the full source code for the Processing sketch is shown in Source Code – Processing Sketch on page 140.

Source Code – Arduino Sketch

This is the code for the Arduino sketch that implements the RCDIO protocol.

```
#include <SPI.h>
#include <Ethernet.h>

const byte mac[] = { 0x00, 0xC3, 0xA2, 0xE6, 0x3D, 0x57 };
byte ip[] = { 192, 168, 0, 99 };

EthernetServer myServer(80);

struct cmd {
  char reserved[3];
  byte opcode;
  byte length;
};

void setup() {
  //D53 on the Arduino Mega must be an output.
  pinMode(53, OUTPUT);

  Serial.begin(9600);
  while (!Serial);

  Serial.println("Establishing network connection...");
  Ethernet.begin((uint8_t*)mac, ip);
  Serial.print("IP Address: ");
  Serial.println(Ethernet.localIP());

  myServer.begin();
}

void loop() {
  cmd myCmd;
  EthernetClient client = myServer.available();
  if (client) {
    while (client.connected()) {
      if (client.available() >= 5) {
        client.read((uint8_t*)&myCmd, sizeof(myCmd));
        break;
      }
    }

    if (
        (myCmd.reserved[0] == 'R') &&
        (myCmd.reserved[1] == 'C') &&
```

```
                (myCmd.reserved[2] == 'D')
        ) {
            int tmp;
            switch (myCmd.opcode) {
              case 1:
                sendInfo(&client);
                break;
              case 2:
                while (client.available() < 2);
                pinMode(client.read(), client.read());
                sendResponse(&client);
                break;
              case 3:
                while (client.available() < 2);
                digitalWrite(client.read(), client.read());
                sendResponse(&client);
                break;
              case 4:
                while (client.available() < 1);
                tmp = digitalRead(client.read());
                sendResponse(&client, (byte)tmp);
                break;
              case 5:
                while (client.available() < 2);
                analogWrite(client.read(), client.read());
                sendResponse(&client);
                break;
              case 6:
                while (client.available() < 1);
                tmp = analogRead(client.read());
                sendResponse(&client, (unsigned int)tmp);
                break;
            }
        }
      client.stop();
    }
}

void sendInfo(EthernetClient* client) {
  char message[] = "RCDIO Version 0.1 - Arduino Uno";
  cmd response;
  response.reserved[0] = 'R';
  response.reserved[1] = 'C';
  response.reserved[2] = 'D';
  response.opcode = 0;
  response.length = strlen(message);
  client->write((uint8_t*)&response, sizeof(response));
  client->write(message);
}

void sendResponse(EthernetClient* client) {
  cmd response;
  response.reserved[0] = 'R';
  response.reserved[1] = 'C';
  response.reserved[2] = 'D';
  response.opcode = 0;
  response.length = 0;
  client->write((uint8_t*)&response, sizeof(response));
}

void sendResponse(EthernetClient* client, byte data) {
  cmd response;
  response.reserved[0] = 'R';
  response.reserved[1] = 'C';
  response.reserved[2] = 'D';
  response.opcode = 0;
  response.length = 1;
  client->write((uint8_t*)&response, sizeof(response));
  client->write(data);
```

```
}

void sendResponse(EthernetClient* client, unsigned int data) {
  cmd response;
  response.reserved[0] = 'R';
  response.reserved[1] = 'C';
  response.reserved[2] = 'D';
  response.opcode = 0;
  response.length = 2;
  client->write((uint8_t*)&response, sizeof(response));
  client->write((byte)(data & 0x0000FFFF));
  client->write((byte)(data >> 8));
}
```

Source Code – Processing Sketch

The following Processing sketch contains six functions that represent the six available commands in the RCDIO protocol. For initial testing, the lines in the sketch's *setup()* function tell the Arduino to make pin 2 an *output* and then bring it *high*. This will turn on a light-emitting diode (LED) if you connect one through a 220Ω resistor.

```
import processing.net.*;

public class cmd {
  char[] reserved = new char[3];
  byte opcode;
  byte oplength;

  public cmd() {
  }

  public cmd(byte[] fromByteArray) {
    reserved[0] = (char)fromByteArray[0];
    reserved[1] = (char)fromByteArray[1];
    reserved[2] = (char)fromByteArray[2];
    opcode = fromByteArray[3];
    oplength = fromByteArray[4];
  }

  public byte[] toByteArray() {
    byte[] result = new byte[5];
    result[0] = (byte)reserved[0];
    result[1] = (byte)reserved[1];
    result[2] = (byte)reserved[2];
    result[3] = opcode;
    result[4] = oplength;
    return result;
  }
}

final int HIGH = 1;
final int LOW = 0;
final int OUTPUT = 1;
final int INPUT = 0;
final String serverIP = "192.168.0.99";

void setup() {
  pinMode(2, OUTPUT);
  digitalWrite(2, HIGH);
}

String getInfo() {
  cmd myCmd = new cmd();
```

```
  myCmd.reserved[0] = 'R';
  myCmd.reserved[1] = 'C';
  myCmd.reserved[2] = 'D';
  myCmd.opcode = 1;
  myCmd.oplength = 0;

  Client c = new Client(this, serverIP, 80);
  c.write(myCmd.toByteArray());
  while (c.available() < 5);
  byte[] hdr = new byte[5];
  c.readBytes(hdr);
  cmd response = new cmd(hdr);
  if (response.opcode == 0) {
    while (c.available() < response.oplength);
    byte[] buf = new byte[256];
    c.readBytes(buf);
    String data = new String(buf);
    c.stop();
    return data;
  }

  c.stop();
  return "";
}

int analogRead(int pin) {
  int result = -1;
  cmd myCmd = new cmd();
  myCmd.reserved[0] = 'R';
  myCmd.reserved[1] = 'C';
  myCmd.reserved[2] = 'D';
  myCmd.opcode = 6;
  myCmd.oplength = 1;

  Client c = new Client(this, serverIP, 80);
  c.write(myCmd.toByteArray());
  c.write((byte)pin);

  while (c.available() < 5);
  byte[] hdr = new byte[5];
  for (int i=0; i<5; i++)
    hdr[i] = (byte)c.read();
  cmd response = new cmd(hdr);
  if (response.oplength > 0) {
    while (c.available() < response.oplength);
    byte tmp1 = (byte)c.read();
    byte tmp2 = (byte)c.read();
    result = 0 + tmp1 + (tmp2 * 256);
  }

  c.stop();
  return result;
}

void analogWrite(int pin, int value) {
  cmd myCmd = new cmd();
  myCmd.reserved[0] = 'R';
  myCmd.reserved[1] = 'C';
  myCmd.reserved[2] = 'D';
  myCmd.opcode = 5;
  myCmd.oplength = 2;

  Client c = new Client(this, serverIP, 80);
  c.write(myCmd.toByteArray());
  c.write((byte)pin);
  c.write((byte)value);
  while (c.available() < 5);
  byte[] hdr = new byte[5];
  c.readBytes(hdr);
```

```
    cmd response = new cmd(hdr);

    c.stop();
}

int digitalRead(int pin) {
    int result = -1;
    cmd myCmd = new cmd();
    myCmd.reserved[0] = 'R';
    myCmd.reserved[1] = 'C';
    myCmd.reserved[2] = 'D';
    myCmd.opcode = 4;
    myCmd.oplength = 1;

    Client c = new Client(this, serverIP, 80);
    c.write(myCmd.toByteArray());
    c.write((byte)pin);
    while (c.available() < 5);
    byte[] hdr = new byte[5];
    c.readBytes(hdr);
    cmd response = new cmd(hdr);
    if (response.oplength > 0) {
      while (c.available() < 1);
      result = 0 + c.read();
    }

    c.stop();
    return result;
}

void digitalWrite(int pin, int mode) {
    cmd myCmd = new cmd();
    myCmd.reserved[0] = 'R';
    myCmd.reserved[1] = 'C';
    myCmd.reserved[2] = 'D';
    myCmd.opcode = 3;
    myCmd.oplength = 2;

    Client c = new Client(this, serverIP, 80);
    c.write(myCmd.toByteArray());
    c.write((byte)pin);
    if (mode == HIGH)
      c.write((byte)1);
    else
      c.write((byte)0);
    while (c.available() < 5);
    byte[] hdr = new byte[5];
    c.readBytes(hdr);
    cmd response = new cmd(hdr);

    c.stop();
}

void pinMode(int pin, int mode) {
    cmd myCmd = new cmd();
    myCmd.reserved[0] = 'R';
    myCmd.reserved[1] = 'C';
    myCmd.reserved[2] = 'D';
    myCmd.opcode = 2;
    myCmd.oplength = 2;

    Client c = new Client(this, serverIP, 80);
    c.write(myCmd.toByteArray());
    c.write((byte)pin);
    if (mode == HIGH)
      c.write((byte)1);
    else
      c.write((byte)0);
```

```
  while (c.available() < 5);
  byte[] hdr = new byte[5];
  c.readBytes(hdr);
  cmd response = new cmd(hdr);
  if (response.opcode != 0)
    println("OK");

  c.stop();
}
```

Appendix A – Hypertext Transfer Protocol – HTTP/1.0

This specification of the hypertext transfer protocol is adapted from the version published by the HTTP Working Group in 1996. Certain sections have been shortened or omitted, and readers are encouraged to refer to the full specification at http://www.w3.org/Protocols/HTTP/1.0/spec.html

Abstract

The Hypertext Transfer Protocol (HTTP) is an application-level protocol with the lightness and speed necessary for distributed, collaborative, hypermedia information systems. It is a generic, stateless, object-oriented protocol which can be used for many tasks, such as name servers and distributed object management systems, through extension of its request methods (commands). A feature of HTTP is the typing of data representation, allowing systems to be built independently of the data being transferred.

HTTP has been in use by the World-Wide Web global information initiative since 1990. This specification reflects common usage of the protocol referred to as "HTTP/1.0".

Table of Contents

1. Introduction

1.1 Purpose

The Hypertext Transfer Protocol (HTTP) is an application-level protocol with the lightness and speed necessary for distributed, collaborative, hypermedia information systems. HTTP has been in use by the World-Wide Web global information initiative since 1990. This specification reflects common usage of the protocol referred to as "HTTP/1.0". This specification describes the features that seem to be consistently implemented in most HTTP/1.0 clients and servers. The specification is split into two sections. Those features of HTTP for which implementations are usually consistent are described in the main body of this document. Those features which have few or inconsistent implementations are listed in Appendix A.D.

Practical information systems require more functionality than simple retrieval, including search, front-end update, and annotation. HTTP allows an open-ended set of methods to be used to indicate the purpose of a request. It builds on the discipline of reference provided by the Uniform Resource Identifier (URI), as a location (URL) or name (URN), for indicating the resource on which a method is to be applied. Messages are passed in a format similar to that used by Internet Mail and the Multipurpose Internet Mail Extensions (MIME).

HTTP is also used as a generic protocol for communication between user agents and proxies/gateways to other Internet protocols, such as SMTP, NNTP, FTP, Gopher, and WAIS, allowing basic hypermedia access to resources available from diverse applications and simplifying the implementation of user agents.

1.2 Terminology

connection
> A transport layer virtual circuit established between two application programs for the purpose of communication.

message
> The basic unit of HTTP communication, consisting of a structured sequence of octets matching the syntax defined in Section 4 and transmitted via the connection.

request
> An HTTP request message (as defined in Section 5).

response
> An HTTP response message (as defined in Section 6).

resource

    A network data object or service which can be identified by a URI (Section 3.2).

entity

    A particular representation or rendition of a data resource, or reply from a service resource, that may be enclosed within a request or response message. An entity consists of metainformation in the form of entity headers and content in the form of an entity body.

client

    An application program that establishes connections for the purpose of sending requests.

user agent

    The client which initiates a request. These are often browsers, editors, spiders (web-traversing robots), or other end user tools.

server

    An application program that accepts connections in order to service requests by sending back responses.

origin server

    The server on which a given resource resides or is to be created.

proxy

    An intermediary program which acts as both a server and a client for the purpose of making requests on behalf of other clients. Requests are serviced internally or by passing them, with possible translation, on to other servers. A proxy must interpret and, if necessary, rewrite a request message before forwarding it. Proxies are often used as client-side portals through network firewalls and as helper applications for handling requests via protocols not implemented by the user agent.

gateway

    A server which acts as an intermediary for some other server. Unlike a proxy, a gateway receives requests as if it were the origin server for the requested resource; the requesting client may not be aware that it is communicating with a gateway. Gateways are often used as server-side portals through network firewalls and as protocol translators for access to resources stored on non-HTTP systems.

tunnel

    A tunnel is an intermediary program which is acting as a blind relay between two connections. Once active, a tunnel is not considered a party to the HTTP communication, though the tunnel may have been initiated by an HTTP request. The tunnel ceases to exist when both ends of the relayed connections are closed. Tunnels are used when a portal is necessary and the intermediary cannot, or should not, interpret the relayed communication.

cache

A program's local store of response messages and the subsystem that controls its message storage, retrieval, and deletion. A cache stores cachable responses in order to reduce the response time and network bandwidth consumption on future, equivalent requests. Any client or server may include a cache, though a cache cannot be used by a server while it is acting as a tunnel.

Any given program may be capable of being both a client and a server; our use of these terms refers only to the role being performed by the program for a particular connection, rather than to the program's capabilities in general. Likewise, any server may act as an origin server, proxy, gateway, or tunnel, switching behavior based on the nature of each request.

1.3 Overall Operation

The HTTP protocol is based on a request/response paradigm. A client establishes a connection with a server and sends a request to the server in the form of a request method, URI, and protocol version, followed by a MIME-like message containing request modifiers, client information, and possible body content. The server responds with a status line, including the message's protocol version and a success or error code, followed by a MIME-like message containing server information, entity metainformation, and possible body content.

Most HTTP communication is initiated by a user agent and consists of a request to be applied to a resource on some origin server. In the simplest case, this may be accomplished via a single connection (v) between the user agent (UA) and the origin server (O).

```
request chain ---------------------------->
UA -------------------v------------------- O
<-------------------------- response chain
```

A more complicated situation occurs when one or more intermediaries are present in the request/response chain. There are three common forms of intermediary: proxy, gateway, and tunnel. A proxy is a forwarding agent, receiving requests for a URI in its absolute form, rewriting all or parts of the message, and forwarding the reformatted request toward the server identified by the URI. A gateway is a receiving agent, acting as a layer above some other server(s) and, if necessary, translating the requests to the underlying server's protocol. A tunnel acts as a relay point between two connections without changing the messages; tunnels are used when the communication needs to pass through an intermediary (such as a firewall) even when the intermediary cannot understand the contents of the messages.

On the Internet, HTTP communication generally takes place over TCP/IP connections. The default port is TCP 80, but other ports can be used. This does not preclude HTTP from being implemented on top of any other protocol on the Internet, or on other networks. HTTP only presumes a reliable transport; any protocol that provides such guarantees can be used, and the mapping of the HTTP/1.0 request and response structures onto the transport data units of the protocol in question is outside the scope of this specification.

Except for experimental applications, current practice requires that the connection be established by the client prior to each request and closed by the server after sending the response. Both clients and servers should be aware that either party may close the connection prematurely, due to user action, automated time-out, or program failure, and should handle such closing in a predictable fashion. In any case, the closing of the connection by either or both parties always terminates the current request, regardless of its status.

1.4 HTTP and MIME

HTTP/1.0 uses many of the constructs defined for MIME, as defined in RFC 1521. Appendix A.C describes the ways in which the context of HTTP allows for different use of Internet Media Types than is typically found in Internet mail, and gives the rationale for those differences.

2. Notational Conventions and Generic Grammar

2.1 Augmented BNF

All of the mechanisms specified in this document are described in both prose and an augmented Backus-Naur Form (BNF) similar to that used by RFC 822. Implementors will need to be familiar with the notation in order to understand this specification.

2.2 Basic Rules

The following rules are used throughout this specification to describe basic parsing constructs.

```
OCTET          = <any 8-bit sequence of data>
CHAR           = <any US-ASCII character (octets 0 - 127)>
UPALPHA        = <any US-ASCII uppercase letter "A".."Z">
LOALPHA        = <any US-ASCII lowercase letter "a".."z">
ALPHA          = UPALPHA | LOALPHA
DIGIT          = <any US-ASCII digit "0".."9">
CTL            = <any US-ASCII control character
                 (octets 0 - 31) and DEL (127)>
CR             = <US-ASCII CR, carriage return (13)>
LF             = <US-ASCII LF, linefeed (10)>
SP             = <US-ASCII SP, space (32)>
HT             = <US-ASCII HT, horizontal-tab (9)>
<">            = <US-ASCII double-quote mark (34)>
```

HTTP/1.0 defines the octet sequence CR LF as the end-of-line marker for all protocol elements except the Entity-Body (see Appendix A.B for tolerant applications). The end-of-line marker within an Entity-Body is defined by its associated media type, as described in Section 3.6.

```
CRLF          = CR LF
```

HTTP/1.0 headers may be folded onto multiple lines if each continuation line begins with a space or horizontal tab. All linear whitespace, including folding, has the same semantics as SP.

```
LWS           = [CRLF] 1*( SP | HT )
```

However, folding of header lines is not expected by some applications, and should not be generated by HTTP/1.0 applications.

The TEXT rule is only used for descriptive field contents and values that are not intended to be interpreted by the message parser. Words of *TEXT may contain octets from character sets other than US-ASCII.

```
TEXT          = <any OCTET except CTLs,
                but including LWS>
```

Recipients of header field TEXT containing octets outside the US-ASCII character set may assume that they represent ISO-8859-1 characters.

Hexadecimal numeric characters are used in several protocol elements.

```
HEX           = "A" | "B" | "C" | "D" | "E" | "F"
              | "a" | "b" | "c" | "d" | "e" | "f" | DIGIT
```

Many HTTP/1.0 header field values consist of words separated by LWS or special characters. These special characters must be in a quoted string to be used within a parameter value.

```
word          = token | quoted-string
token         = 1*<any CHAR except CTLs or tspecials>
tspecials     = "(" | ")" | "<" | ">" | "@"
              | "," | ";" | ":" | "\" | <">
              | "/" | "[" | "]" | "?" | "="
              | "{" | "}" | SP | HT
```

Comments may be included in some HTTP header fields by surrounding the comment text with parentheses. Comments are only allowed in fields containing "comment" as part of their field value definition. In all other fields, parentheses are considered part of the field value.

```
comment       = "(" *( ctext | comment ) ")"
ctext         = <any TEXT excluding "(" and ")">
```

A string of text is parsed as a single word if it is quoted using double-quote marks.

```
quoted-string  = ( <"> *(qdtext) <"> )
dtext          = <any CHAR except <"> and CTLs,
                 but including LWS>
```

Single-character quoting using the backslash ("\") character is not permitted in HTTP/1.0.

3. Protocol Parameters

3.1 HTTP Version

HTTP uses a "<major>.<minor>" numbering scheme to indicate versions of the protocol. The protocol versioning policy is intended to allow the sender to indicate the format of a message and its capacity for understanding further HTTP communication, rather than the features obtained via that communication. No change is made to the version number for the addition of message components which do not affect communication behavior or which only add to extensible field values. The <minor> number is incremented when the changes made to the protocol add features which do not change the general message parsing algorithm, but which may add to the message semantics and imply additional capabilities of the sender. The <major> number is incremented when the format of a message within the protocol is changed.

The version of an HTTP message is indicated by an HTTP-Version field in the first line of the message. If the protocol version is not specified, the recipient must assume that the message is in the simple HTTP/0.9 format.

```
HTTP-Version  = "HTTP" "/" 1*DIGIT "." 1*DIGIT
```

Note that the major and minor numbers should be treated as separate integers and that each may be incremented higher than a single digit. Thus, HTTP/2.4 is a lower version than HTTP/2.13, which in turn is lower than HTTP/12.3. Leading zeros should be ignored by recipients and never generated by senders.

This document defines both the 0.9 and 1.0 versions of the HTTP protocol. Applications sending Full-Request or Full-Response messages, as defined by this specification, must include an HTTP-Version of "HTTP/1.0".

HTTP/1.0 servers must:

- recognize the format of the Request-Line for HTTP/0.9 and HTTP/1.0 requests;
- understand any valid request in the format of HTTP/0.9 or HTTP/1.0;
- respond appropriately with a message in the same protocol version used by the client.

HTTP/1.0 clients must:

- recognize the format of the Status-Line for HTTP/1.0 responses;
- understand any valid response in the format of HTTP/0.9 or HTTP/1.0.

Proxy and gateway applications must be careful in forwarding requests that are received in a format different than that of the application's native HTTP version. Since the protocol version indicates the protocol capability of the sender, a proxy/gateway must never send a message with a version indicator which is greater than its native version; if a higher version request is received, the proxy/gateway must either downgrade the request version or respond with an error. Requests with a version lower than that of the application's native format may be upgraded before being forwarded; the proxy/gateway's response to that request must follow the server requirements listed above.

3.2 Uniform Resource Identifiers

URIs have been known by many names: WWW addresses, Universal Document Identifiers, Universal Resource Identifiers, and finally the combination of Uniform Resource Locators (URL) and Names (URN) As far as HTTP is concerned, Uniform Resource Identifiers are simply formatted strings which identify--via name, location, or any other characteristic--a network resource.

3.2.1 General Syntax

URIs in HTTP can be represented in absolute form or relative to some known base URI, depending upon the context of their use. The two forms are differentiated by the fact that absolute URIs always begin with a scheme name followed by a colon.

```
URI           = ( absoluteURI | relativeURI ) [ "#" fragment ]

absoluteURI   = scheme ":" *( uchar | reserved )

relativeURI   = net_path | abs_path | rel_path

net_path      = "//" net_loc [ abs_path ]
abs_path      = "/" rel_path
rel_path      = [ path ] [ ";" params ] [ "?" query ]
```

```
path           = fsegment *( "/" segment )
fsegment       = 1*pchar
segment        = *pchar

params         = param *( ";" param )
param          = *( pchar | "/" )

scheme         = 1*( ALPHA | DIGIT | "+" | "-" | "." )
net_loc        = *( pchar | ";" | "?" )
query          = *( uchar | reserved )
fragment       = *( uchar | reserved )

pchar          = uchar | ":" | "@" | "&" | "=" | "+"
uchar          = unreserved | escape
unreserved     = ALPHA | DIGIT | safe | extra | national

escape         = "%" HEX HEX
reserved       = ";" | "/" | "?" | ":" | "@" | "&" | "=" | "+"
extra          = "!" | "*" | "'" | "(" | ")" | ","
safe           = "$" | "-" | "_" | "."
unsafe         = CTL | SP | <"> | "#" | "%" | "<" | ">"
national       = <any OCTET excluding ALPHA, DIGIT,
                  reserved, extra, safe, and unsafe>
```

For definitive information on URL syntax and semantics, see RFC 1738 and RFC 1808. The BNF above includes national characters not allowed in valid URLs as specified by RFC 1738, since HTTP servers are not restricted in the set of unreserved characters allowed to represent the rel_path part of addresses, and HTTP proxies may receive requests for URIs not defined by RFC 1738.

3.2.2 http URL

The "http" scheme is used to locate network resources via the HTTP protocol. This section defines the scheme-specific syntax and semantics for http URLs.

```
http_URL       = "http:" "//" host [ ":" port ] [ abs_path ]

host           = <A legal Internet host domain name
                  or IP address (in dotted-decimal form),
                  as defined by Section 2.1 of RFC 1123>

port           = *DIGIT
```

If the port is empty or not given, port 80 is assumed. The semantics are that the identified resource is located at the server listening for TCP connections on that port of that host, and the Request-URI for the resource is abs_path. If the abs_path is not present in the URL, it must be given as "/" when used as a Request-URI (Section 5.1.2).

The canonical form for "http" URLs is obtained by converting any UPALPHA characters in host to their LOALPHA equivalent (hostnames are case-insensitive), eliding the [":" port] if the port is 80, and replacing an empty abs_path with "/".

3.3 Date/Time Formats

HTTP/1.0 applications have historically allowed three different formats for the representation of date/time stamps:

```
Sun, 06 Nov 1994 08:49:37 GMT    ; RFC 822, updated by RFC 1123
Sunday, 06-Nov-94 08:49:37 GMT   ; RFC 850, obsoleted by RFC 1036
Sun Nov  6 08:49:37 1994         ; ANSI C's asctime() format
```

The first format is preferred as an Internet standard and represents a fixed-length subset of that defined by RFC 1123 (an update to RFC 822). The second format is in common use, but is based on the obsolete RFC 850 date format and lacks a four-digit year. HTTP/1.0 clients and servers that parse the date value should accept all three formats, though they must never generate the third (asctime) format.

All HTTP/1.0 date/time stamps must be represented in Universal Time (UT), also known as Greenwich Mean Time (GMT), without exception. This is indicated in the first two formats by the inclusion of "GMT" as the three-letter abbreviation for time zone, and should be assumed when reading the asctime format.

```
HTTP-date      = rfc1123-date | rfc850-date | asctime-date

rfc1123-date   = wkday "," SP date1 SP time SP "GMT"
rfc850-date    = weekday "," SP date2 SP time SP "GMT"
asctime-date   = wkday SP date3 SP time SP 4DIGIT

date1          = 2DIGIT SP month SP 4DIGIT
                 ; day month year (e.g., 02 Jun 1982)
date2          = 2DIGIT "-" month "-" 2DIGIT
                 ; day-month-year (e.g., 02-Jun-82)
date3          = month SP ( 2DIGIT | ( SP 1DIGIT ))
                 ; month day (e.g., Jun  2)

time           = 2DIGIT ":" 2DIGIT ":" 2DIGIT
                 ; 00:00:00 - 23:59:59

wkday          = "Mon" | "Tue" | "Wed"
               | "Thu" | "Fri" | "Sat" | "Sun"

weekday        = "Monday" | "Tuesday" | "Wednesday"
               | "Thursday" | "Friday" | "Saturday" | "Sunday"

month          = "Jan" | "Feb" | "Mar" | "Apr"
               | "May" | "Jun" | "Jul" | "Aug"
               | "Sep" | "Oct" | "Nov" | "Dec"
```

3.4 Character Sets

HTTP uses the same definition of the term "character set" as that described for MIME:

The term "character set" is used in this document to refer to a method used with one or more tables to convert a sequence of octets into a sequence of characters. Note that unconditional conversion in the other direction is not required, in that not all characters may be available in a given character set and a character set may provide more than one sequence of octets to represent a particular character. This definition is intended to allow various kinds of character encodings, from simple single-table mappings such as US-ASCII to complex table switching methods such as those that use ISO 2022's techniques. However, the definition associated with a MIME character set name must fully specify the mapping to be performed from octets to characters. In particular, use of external profiling information to determine the exact mapping is not permitted.

Note: This use of the term "character set" is more commonly referred to as a "character encoding." However, since HTTP and MIME share the same registry, it is important that the terminology also be shared.

HTTP character sets are identified by case-insensitive tokens. The complete set of tokens are defined by the IANA Character Set registry. However, because that registry does not define a single, consistent token for each character set, we define here the preferred names for those character sets most likely to be used with HTTP entities. These character sets include those registered by RFC 1521 -- the US-ASCII and ISO-8859 character sets -- and other names specifically recommended for use within MIME charset parameters.

```
charset = "US-ASCII"
        | "ISO-8859-1" | "ISO-8859-2" | "ISO-8859-3"
        | "ISO-8859-4" | "ISO-8859-5" | "ISO-8859-6"
        | "ISO-8859-7" | "ISO-8859-8" | "ISO-8859-9"
        | "ISO-2022-JP" | "ISO-2022-JP-2" | "ISO-2022-KR"
        | "UNICODE-1-1" | "UNICODE-1-1-UTF-7" | "UNICODE-1-1-UTF-8"
        | token
```

Although HTTP allows an arbitrary token to be used as a charset value, any token that has a predefined value within the IANA Character Set registry must represent the character set defined by that registry. Applications should limit their use of character sets to those defined by the IANA registry.

The character set of an entity body should be labelled as the lowest common denominator of the character codes used within that body, with the exception that no label is preferred over the labels US-ASCII or ISO-8859-1.

3.5 Content Codings

Content coding values are used to indicate an encoding transformation that has been applied to a resource. Content codings are primarily used to allow a document to be compressed or encrypted without losing the identity of its underlying media type. Typically, the resource is stored in this encoding and only decoded before rendering or analogous usage.

```
content-coding        = "x-gzip" | "x-compress" | token
```

Note: For future compatibility, HTTP/1.0 applications should consider "gzip" and "compress" to be equivalent to "x-gzip" and "x-compress", respectively.

All content-coding values are case-insensitive. HTTP/1.0 uses content-coding values in the Content-Encoding (Section 10.3) header field. Although the value describes the content-coding, what is more important is that it indicates what decoding mechanism will be required to remove the encoding. Note that a single program may be capable of decoding multiple content-coding formats.

3.6 Media Types

HTTP uses Internet Media Types in the Content-Type header field (Section 10.5) in order to provide open and extensible data typing.

```
media-type      = type "/" subtype *( ";" parameter )
type            = token
subtype         = token
```

Parameters may follow the type/subtype in the form of attribute/value pairs.

```
parameter       = attribute "=" value
attribute       = token
value           = token | quoted-string
```

The type, subtype, and parameter attribute names are case-insensitive. Parameter values may or may not be case-sensitive, depending on the semantics of the parameter name. LWS must not be generated between the type and subtype, nor between an attribute and its value. Upon receipt of a media type with an unrecognized parameter, a user agent should treat the media type as if the unrecognized parameter and its value were not present.

Some older HTTP applications do not recognize media type parameters. HTTP/1.0 applications should only use media type parameters when they are necessary to define the content of a message.

Media-type values are registered with the Internet Assigned Number Authority (IANA). The media type registration process is outlined in RFC 1590. Use of non-registered media types is discouraged.

3.6.1 Canonicalization and Text Defaults

Internet media types are registered with a canonical form. In general, an Entity-Body transferred via HTTP must be represented in the appropriate canonical form prior to its transmission. If the body has been encoded with a Content-Encoding, the underlying data should be in canonical form prior to being encoded.

Media subtypes of the "text" type use CRLF as the text line break when in canonical form. However, HTTP allows the transport of text media with plain CR or LF alone representing a line break when used consistently within the Entity-Body. HTTP applications must accept CRLF, bare CR, and bare LF as being representative of a line break in text media received via HTTP.

In addition, if the text media is represented in a character set that does not use octets 13 and 10 for CR and LF respectively, as is the case for some multi-byte character sets, HTTP allows the use of whatever octet sequences are defined by that character set to represent the equivalent of CR and LF for line breaks. This flexibility regarding line breaks applies only to text media in the Entity-Body; a bare CR or LF should not be substituted for CRLF within any of the HTTP control structures (such as header fields and multipart boundaries).

The "charset" parameter is used with some media types to define the character set (Section 3.4) of the data. When no explicit charset parameter is provided by the sender, media subtypes of the "text" type are defined to have a default charset value of "ISO-8859-1" when received via HTTP. Data in character sets other than "ISO-8859-1" or its subsets must be labelled with an appropriate charset value in order to be consistently interpreted by the recipient.

3.6.2 Multipart Types

MIME provides for a number of "multipart" types -- encapsulations of several entities within a single message's Entity-Body. The multipart types registered by IANA do not have any special meaning for HTTP/1.0, though user agents may need to understand each type in order to correctly interpret the purpose of each body-part. An HTTP user agent should follow the same or similar behavior as a MIME user agent does upon receipt of a multipart type. HTTP servers should not assume that all HTTP clients are prepared to handle multipart types.

All multipart types share a common syntax and must include a boundary parameter as part of the media type value. The message body is itself a protocol element and must therefore use only CRLF to represent line breaks between body-parts. Multipart body-parts may contain HTTP header fields which are significant to the meaning of that part.

3.7 Product Tokens

Product tokens are used to allow communicating applications to identify themselves via a simple product token, with an optional slash and version designator. Most fields using product tokens also allow subproducts which form a significant part of the application to be listed, separated by whitespace. By convention, the products are listed in order of their significance for identifying the application.

```
product          = token ["/" product-version]
product-version = token
```

Examples:

```
User-Agent: CERN-LineMode/2.15 libwww/2.17b3

Server: Apache/0.8.4
```

Product tokens should be short and to the point -- use of them for advertizing or other non-essential information is explicitly forbidden. Although any token character may appear in a product-version, this token should only be used for a version identifier (i.e., successive versions of the same product should only differ in the product-version portion of the product value).

4. HTTP Message

4.1 Message Types

HTTP messages consist of requests from client to server and responses from server to client.

```
HTTP-message   = Simple-Request          ; HTTP/0.9 messages
               | Simple-Response
               | Full-Request            ; HTTP/1.0 messages
               | Full-Response
```

Full-Request and Full-Response use the generic message format of RFC 822 for transferring entities. Both messages may include optional header fields (also known as "headers") and an entity body. The entity body is separated from the headers by a null line (i.e., a line with nothing preceding the CRLF).

```
Full-Request   = Request-Line            ; Section 5.1
                 *( General-Header        ; Section 4.3
                  | Request-Header        ; Section 5.2
                  | Entity-Header )       ; Section 7.1
                 CRLF
                 [ Entity-Body ]          ; Section 7.2

Full-Response  = Status-Line             ; Section 6.1
                 *( General-Header        ; Section 4.3
                  | Response-Header       ; Section 6.2
                  | Entity-Header )       ; Section 7.1
                 CRLF
                 [ Entity-Body ]          ; Section 7.2
```

Simple-Request and Simple-Response do not allow the use of any header information and are limited to a single request method (GET).

```
Simple-Request  = "GET" SP Request-URI CRLF

Simple-Response = [ Entity-Body ]
```

Use of the Simple-Request format is discouraged because it prevents the server from identifying the media type of the returned entity.

4.2 Message Headers

HTTP header fields, which include General-Header (Section 4.3), Request-Header (Section 5.2), Response-Header (Section 6.2), and Entity-Header (Section 7.1) fields, follow the same generic format as that given in Section 3.1 of RFC 822. Each header field consists of a name followed immediately by a colon (":"), a single space (SP) character, and the field value. Field names are case-insensitive. Header fields can be extended over multiple lines by preceding each extra line with at least one SP or HT, though this is not recommended.

```
HTTP-header    = field-name ":" [ field-value ] CRLF

field-name     = token
field-value    = *( field-content | LWS )

field-content  = <the OCTETs making up the field-value
                  and consisting of either *TEXT or combinations
                  of token, tspecials, and quoted-string>
```

The order in which header fields are received is not significant. However, it is "good practice" to send General-Header fields first, followed by Request-Header or Response-Header fields prior to the Entity-Header fields.

Multiple HTTP-header fields with the same field-name may be present in a message if and only if the entire field-value for that header field is defined as a comma-separated list [i.e., #(values)]. It must be possible to combine the multiple header fields into one "field-name: field-value" pair, without changing the semantics of the message, by appending each subsequent field-value to the first, each separated by a comma.

4.3 General Header Fields

There are a few header fields which have general applicability for both request and response messages, but which do not apply to the entity being transferred. These headers apply only to the message being transmitted.

```
General-Header = Date                   ; Section 10.6
               | Pragma                 ; Section 10.12
```

General header field names can be extended reliably only in combination with a change in the protocol version. However, new or experimental header fields may be given the semantics of general header fields if all parties in the communication recognize them to be general header fields. Unrecognized header fields are treated as Entity-Header fields.

5. Request

A request message from a client to a server includes, within the first line of that message, the method to be applied to the resource, the identifier of the resource, and the protocol version in use. For backwards compatibility with the more limited HTTP/0.9 protocol, there are two valid formats for an HTTP request:

```
Request         = Simple-Request | Full-Request

Simple-Request = "GET" SP Request-URI CRLF

Full-Request    = Request-Line             ; Section 5.1
                  *( General-Header         ; Section 4.3
                   | Request-Header         ; Section 5.2
                   | Entity-Header )        ; Section 7.1
                  CRLF
                  [ Entity-Body ]           ; Section 7.2
```

If an HTTP/1.0 server receives a Simple-Request, it must respond with an HTTP/0.9 Simple-Response. An HTTP/1.0 client capable of receiving a Full-Response should never generate a Simple-Request.

5.1 Request-Line

The Request-Line begins with a method token, followed by the Request-URI and the protocol version, and ending with CRLF. The elements are separated by SP characters. No CR or LF are allowed except in the final CRLF sequence.

```
Request-Line    = Method SP Request-URI SP HTTP-Version CRLF
```

5.1.1 Method

The Method token indicates the method to be performed on the resource identified by the Request-URI. The method is case-sensitive.

```
Method          = "GET"                    ; Section 8.1
                  | "HEAD"                  ; Section 8.2
                  | "POST"                  ; Section 8.3
                  | extension-method

extension-method = token
```

The list of methods acceptable by a specific resource can change dynamically; the client is notified through the return code of the response if a method is not allowed on a resource. Servers should return the status code 501 (not implemented) if the method is unrecognized or not implemented.

The methods commonly used by HTTP/1.0 applications are fully defined in Section 8.

5.1.2 Request-URI

The Request-URI is a Uniform Resource Identifier (Section 3.2) and identifies the resource upon which to apply the request.

```
Request-URI    = absoluteURI | abs_path
```

The two options for Request-URI are dependent on the nature of the request.

The absoluteURI form is only allowed when the request is being made to a proxy. The proxy is requested to forward the request and return the response. If the request is GET or HEAD and a prior response is cached, the proxy may use the cached message if it passes any restrictions in the Expires header field.

Note that the proxy may forward the request on to another proxy or directly to the server specified by the absoluteURI. In order to avoid request loops, a proxy must be able to recognize all of its server names, including any aliases, local variations, and the numeric IP address. An example Request-Line would be:

```
GET /TheProject.html HTTP/1.0
```

The most common form of Request-URI is that used to identify a resource on an origin server or gateway. In this case, only the absolute path of the URI is transmitted (see Section 3.2.1, abs_path). For example, a client wishing to retrieve the resource above directly from the origin server would create a TCP connection to port 80 of the host "www.w3.org" and send the line:

```
GET /pub/WWW/TheProject.html HTTP/1.0
```

followed by the remainder of the Full-Request. Note that the absolute path cannot be empty; if none is present in the original URI, it must be given as "/" (the server root).

The Request-URI is transmitted as an encoded string, where some characters may be escaped using the "% HEX HEX" encoding defined by RFC 1738. The origin server must decode the Request-URI in order to properly interpret the request.

5.2 Request Header Fields

The request header fields allow the client to pass additional information about the request, and about the client itself, to the server. These fields act as request modifiers, with semantics equivalent to the parameters on a programming language method (procedure) invocation.

```
Request-Header = Authorization          ; Section 10.2
               | From                    ; Section 10.8
               | If-Modified-Since       ; Section 10.9
               | Referer                 ; Section 10.13
               | User-Agent              ; Section 10.15
```

Request-Header field names can be extended reliably only in combination with a change in the protocol version. However, new or experimental header fields may be given the semantics of request header fields if all parties in the communication recognize them to be request header fields. Unrecognized header fields are treated as Entity-Header fields.

6. Response

After receiving and interpreting a request message, a server responds in the form of an HTTP response message.

```
Response        = Simple-Response | Full-Response

Simple-Response = [ Entity-Body ]

Full-Response   = Status-Line              ; Section 6.1
                  *( General-Header         ; Section 4.3
                  |  Response-Header        ; Section 6.2
                  |  Entity-Header )        ; Section 7.1
                  CRLF
                  [ Entity-Body ]           ; Section 7.2
```

A Simple-Response should only be sent in response to an HTTP/0.9 Simple-Request or if the server only supports the more limited HTTP/0.9 protocol. If a client sends an HTTP/1.0 Full-Request and receives a response that does not begin with a Status-Line, it should assume that the response is a Simple-Response and parse it accordingly. Note that the Simple-Response consists only of the entity body and is terminated by the server closing the connection.

6.1 Status-Line

The first line of a Full-Response message is the Status-Line, consisting of the protocol version followed by a numeric status code and its associated textual phrase, with each element separated by SP characters. No CR or LF is allowed except in the final CRLF sequence.

```
Status-Line = HTTP-Version SP Status-Code SP Reason-Phrase CRLF
```

Since a status line always begins with the protocol version and status code

```
"HTTP/" 1*DIGIT "." 1*DIGIT SP 3DIGIT SP
```

(e.g., "HTTP/1.0 200 "), the presence of that expression is sufficient to differentiate a Full-Response from a Simple-Response. Although the Simple-Response format may allow such an expression to occur at the beginning of an entity body, and thus cause a misinterpretation of the message if it was given in response to a Full-Request, most HTTP/0.9 servers are limited to responses of type "text/html" and therefore would never generate such a response.

6.1.1 Status Code and Reason Phrase

The Status-Code element is a 3-digit integer result code of the attempt to understand and satisfy the request. The Reason-Phrase is intended to give a short textual description of the Status-Code. The Status-Code is intended for use by automata and the Reason-Phrase is intended for the human user. The client is not required to examine or display the Reason-Phrase.

The first digit of the Status-Code defines the class of response. The last two digits do not have any categorization role. There are 5 values for the first digit:

•1xx: Informational - Not used, but reserved for future use.

•2xx: Success - The action was successfully received, understood, and accepted.

•3xx: Redirection - Further action must be taken in order to complete the request.

•4xx: Client Error - The request contains bad syntax or cannot be fulfilled.

•5xx: Server Error - The server failed to fulfill an apparently valid request.

HTTP status codes are extensible, but the above codes are the only ones generally recognized in current practice. HTTP applications are not required to understand the meaning of all registered status codes, though such understanding is obviously desirable. However, applications must understand the class of any status code, as indicated by the first digit, and treat any unrecognized response as being equivalent to the x00 status code of that class, with the exception that an unrecognized response must not be cached.

6.2 Response Header Fields

The response header fields allow the server to pass additional information about the response which cannot be placed in the Status-Line. These header fields give information about the server and about further access to the resource identified by the Request-URI.

```
Response-Header = Location            ; Section 10.11
                | Server              ; Section 10.14
                | WWW-Authenticate    ; Section 10.16
```

Response-Header field names can be extended reliably only in combination with a change in the protocol version. However, new or experimental header fields may be given the semantics of response header fields if all parties in the communication recognize them to be response header fields. Unrecognized header fields are treated as Entity-Header fields.

7. Entity

Full-Request and Full-Response messages may transfer an entity within some requests and responses. An entity consists of Entity-Header fields and (usually) an Entity-Body. In this section, both sender and recipient refer to either the client or the server, depending on who sends and who receives the entity.

7.1 Entity Header Fields

Entity-Header fields define optional metainformation about the Entity-Body or, if no body is present, about the resource identified by the request.

```
Entity-Header  = Allow               ; Section 10.1
               | Content-Encoding     ; Section 10.3
               | Content-Length       ; Section 10.4
               | Content-Type         ; Section 10.5
               | Expires              ; Section 10.7
               | Last-Modified        ; Section 10.10
               | extension-header

extension-header = HTTP-header
```

The extension-header mechanism allows additional Entity-Header fields to be defined without changing the protocol, but these fields cannot be assumed to be recognizable by the recipient. Unrecognized header fields should be ignored by the recipient and forwarded by proxies.

7.2 Entity Body

The entity body (if any) sent with an HTTP request or response is in a format and encoding defined by the Entity-Header fields.

```
Entity-Body    = *OCTET
```

An entity body is included with a request message only when the request method calls for one. The presence of an entity body in a request is signaled by the inclusion of a Content-Length header field in the request message headers. HTTP/1.0 requests containing an entity body must include a valid Content-Length header field.

For response messages, whether or not an entity body is included with a message is dependent on both the request method and the response code. All responses to the HEAD request method must not include a body, even though the presence of entity header fields may lead one to believe they do. All 1xx (informational), 204 (no content), and 304 (not modified) responses must not include a body. All other responses must include an entity body or a Content-Length header field defined with a value of zero (0).

7.2.1 Type

When an Entity-Body is included with a message, the data type of that body is determined via the header fields Content-Type and Content-Encoding. These define a two-layer, ordered encoding model:

```
entity-body := Content-Encoding( Content-Type( data ) )
```

A Content-Type specifies the media type of the underlying data. A Content-Encoding may be used to indicate any additional content coding applied to the type, usually for the purpose of data compression, that is a property of the resource requested. The default for the content encoding is none (i.e., the identity function).

Any HTTP/1.0 message containing an entity body should include a Content-Type header field defining the media type of that body. If and only if the media type is not given by a Content-Type header, as is the case for Simple-Response messages, the recipient may attempt to guess the media type via inspection of its content and/or the name extension(s) of the URL used to identify the resource. If the media type remains unknown, the recipient should treat it as type "application/octet-stream".

7.2.2 Length

When an Entity-Body is included with a message, the length of that body may be determined in one of two ways. If a Content-Length header field is present, its value in bytes represents the length of the Entity-Body. Otherwise, the body length is determined by the closing of the connection by the server.

Closing the connection cannot be used to indicate the end of a request body, since it leaves no possibility for the server to send back a response. Therefore, HTTP/1.0 requests containing an entity body must include a valid Content-Length header field. If a request contains an entity body and Content-Length is not specified, and the server does not recognize or cannot calculate the length from other fields, then the server should send a 400 (bad request) response.

8. Method Definitions

The set of common methods for HTTP/1.0 is defined below. Although this set can be expanded, additional methods cannot be assumed to share the same semantics for separately extended clients and servers.

8.1 GET

The GET method means retrieve whatever information (in the form of an entity) is identified by the Request-URI. If the Request-URI refers to a data-producing process, it is the produced data which shall be returned as the entity in the response and not the source text of the process, unless that text happens to be the output of the process.

The semantics of the GET method changes to a "conditional GET" if the request message includes an If-Modified-Since header field. A conditional GET method requests that the identified resource be transferred only if it has been modified since the date given by the If-Modified-Since header, as described in Section 10.9. The conditional GET method is intended to reduce network usage by allowing cached entities to be refreshed without requiring multiple requests or transferring unnecessary data.

8.2 HEAD

The HEAD method is identical to GET except that the server must not return any Entity-Body in the response. The metainformation contained in the HTTP headers in response to a HEAD request should be identical to the information sent in response to a GET request. This method can be used for obtaining metainformation about the resource identified by the Request-URI without transferring the Entity-Body itself. This method is often used for testing hypertext links for validity, accessibility, and recent modification.

There is no "conditional HEAD" request analogous to the conditional GET. If an If-Modified-Since header field is included with a HEAD request, it should be ignored.

8.3 POST

The POST method is used to request that the destination server accept the entity enclosed in the request as a new subordinate of the resource identified by the Request-URI in the Request-Line. POST is designed to allow a uniform method to cover the following functions:

- Annotation of existing resources;
- Posting a message to a bulletin board, newsgroup, mailing list, or similar group of articles;
- Providing a block of data, such as the result of submitting a form [3], to a data-handling process;
- Extending a database through an append operation.

The actual function performed by the POST method is determined by the server and is usually dependent on the Request-URI. The posted entity is subordinate to that URI in the same way that a file is subordinate to a directory containing it, a news article is subordinate to a newsgroup to which it is posted, or a record is subordinate to a database.

A successful POST does not require that the entity be created as a resource on the origin server or made accessible for future reference. That is, the action performed by the POST method might not result in a resource that can be identified by a URI. In this case, either 200 (ok) or 204 (no content) is the appropriate response status, depending on whether or not the response includes an entity that describes the result.

If a resource has been created on the origin server, the response should be 201 (created) and contain an entity (preferably of type "text/html") which describes the status of the request and refers to the new resource.

A valid Content-Length is required on all HTTP/1.0 POST requests. An HTTP/1.0 server should respond with a 400 (bad request) message if it cannot determine the length of the request message's content.

Applications must not cache responses to a POST request because the application has no way of knowing that the server would return an equivalent response on some future request.

9. Status Code Definitions

Each Status-Code is described below, including a description of which method(s) it can follow and any metainformation required in the response.

9.1 Informational 1xx

This class of status code indicates a provisional response, consisting only of the Status-Line and optional headers, and is terminated by an empty line. HTTP/1.0 does not define any 1xx status codes and they are not a valid response to a HTTP/1.0 request. However, they may be useful for experimental applications which are outside the scope of this specification.

9.2 Successful 2xx

This class of status code indicates that the client's request was successfully received, understood, and accepted.

200 OK

The request has succeeded. The information returned with the response is dependent on the method used in the request, as follows:

GET an entity corresponding to the requested resource is sent in the response; HEAD the response must only contain the header information and no Entity-Body; POST an entity describing or containing the result of the action.

201 Created

The request has been fulfilled and resulted in a new resource being created. The newly created resource can be referenced by the URI(s) returned in the entity of the response. The origin server should create the resource before using this Status-Code. If the action cannot be carried out immediately, the server must include in the response body a description of when the resource will be available; otherwise, the server should respond with 202 (accepted).

Of the methods defined by this specification, only POST can create a resource.

202 Accepted

The request has been accepted for processing, but the processing has not been completed. The request may or may not eventually be acted upon, as it may be disallowed when processing actually takes place. There is no facility for re-sending a status code from an asynchronous operation such as this.

The 202 response is intentionally non-committal. Its purpose is to allow a server to accept a request for some other process (perhaps a batch-oriented process that is only run once per day) without requiring that the user agent's connection to the server persist until the process is completed. The entity returned with this response should include an indication of the request's current status and either a pointer to a status monitor or some estimate of when the user can expect the request to be fulfilled.

204 No Content

The server has fulfilled the request but there is no new information to send back. If the client is a user agent, it should not change its document view from that which caused the request to be generated. This response is primarily intended to allow input for scripts or other actions to take place without causing a change to the user agent's active document view. The response may include new metainformation in the form of entity headers, which should apply to the document currently in the user agent's active view.

9.3 Redirection 3xx

This class of status code indicates that further action needs to be taken by the user agent in order to fulfill the request. The action required may be carried out by the user agent without interaction with the user if and only if the method used in the subsequent request is GET or HEAD. A user agent should never automatically redirect a request more than 5 times, since such redirections usually indicate an infinite loop.

300 Multiple Choices

This response code is not directly used by HTTP/1.0 applications, but serves as the default for interpreting the 3xx class of responses.

The requested resource is available at one or more locations. Unless it was a HEAD request, the response should include an entity containing a list of resource characteristics and locations from which the user or user agent can choose the one most appropriate. If the server has a preferred choice, it should include the URL in a Location field; user agents may use this field value for automatic redirection.

301 Moved Permanently

The requested resource has been assigned a new permanent URL and any future references to this resource should be done using that URL. Clients with link editing capabilities should automatically relink references to the Request-URI to the new reference returned by the server, where possible.

The new URL must be given by the Location field in the response. Unless it was a HEAD request, the Entity Body of the response should contain a short note with a hyperlink to the new URL.

If the 301 status code is received in response to a request using the POST method, the user agent must not automatically redirect the request unless it can be confirmed by the user, since this might change the conditions under which the request was issued.

302 Moved Temporarily

The requested resource resides temporarily under a different URL. Since the redirection may be altered on occasion, the client should continue to use the Request-URI for future requests.

The URL must be given by the Location field in the response. Unless it was a HEAD request, the Entity-Body of the response should contain a short note with a hyperlink to the new URI(s).

If the 302 status code is received in response to a request using the POST method, the user agent must not automatically redirect the request unless it can be confirmed by the user, since this might change the conditions under which the request was issued.

304 Not Modified

If the client has performed a conditional GET request and access is allowed, but the document has not been modified since the date and time specified in the If-Modified-Since field, the server must respond with this status code and not send an Entity-Body to the client. Header fields contained in the response should only include information which is relevant to cache managers or which may have changed independently of the entity's Last-Modified date. Examples of relevant header fields include: Date, Server, and Expires. A cache should update its cached entity to reflect any new field values given in the 304 response.

9.4 Client Error 4xx

The 4xx class of status code is intended for cases in which the client seems to have erred. If the client has not completed the request when a 4xx code is received, it should immediately cease sending data to the server. Except when responding to a HEAD request, the server should include an entity containing an explanation of the error situation, and whether it is a temporary or permanent condition. These status codes are applicable to any request method.

Note: If the client is sending data, server implementations on TCP should be careful to ensure that the client acknowledges receipt of the packet(s) containing the response prior to closing the input connection. If the client continues sending data to the server after the close, the server's controller will send a reset packet to the client, which may erase the client's unacknowledged input buffers before they can be read and interpreted by the HTTP application.

400 Bad Request

The request could not be understood by the server due to malformed syntax. The client should not repeat the request without modifications.

401 Unauthorized

The request requires user authentication. The response must include a WWW-Authenticate header field (Section 10.16) containing a challenge applicable to the requested resource. The client may repeat the request with a suitable Authorization header field (Section 10.2). If the request already included Authorization credentials, then the 401 response indicates that authorization has been refused for those credentials. If the 401 response contains the same challenge as the prior response, and the user agent has already attempted authentication at least once, then the user should be presented the entity that was given in the response, since that entity may include relevant diagnostic information. HTTP access authentication is explained in Section 11.

403 Forbidden

The server understood the request, but is refusing to fulfill it. Authorization will not help and the request should not be repeated. If the request method was not HEAD and the server wishes to make public why the request has not been fulfilled, it should describe the reason for the refusal in the entity body. This status code is commonly used when the server does not wish to reveal exactly why the request has been refused, or when no other response is applicable.

404 Not Found

The server has not found anything matching the Request-URI. No indication is given of whether the condition is temporary or permanent. If the server does not wish to make this information available to the client, the status code 403 (forbidden) can be used instead.

9.5 Server Error 5xx

Response status codes beginning with the digit "5" indicate cases in which the server is aware that it has erred or is incapable of performing the request. If the client has not completed the request when a 5xx code is received, it should immediately cease sending data to the server. Except when responding to a HEAD request, the server should include an entity containing an explanation of the error situation, and whether it is a temporary or permanent condition. These response codes are applicable to any request method and there are no required header fields.

500 Internal Server Error

The server encountered an unexpected condition which prevented it from fulfilling the request.

501 Not Implemented

The server does not support the functionality required to fulfill the request. This is the appropriate response when the server does not recognize the request method and is not capable of supporting it for any resource.

502 Bad Gateway

The server, while acting as a gateway or proxy, received an invalid response from the upstream server it accessed in attempting to fulfill the request.

503 Service Unavailable

The server is currently unable to handle the request due to a temporary overloading or maintenance of the server. The implication is that this is a temporary condition which will be alleviated after some delay.

10. Header Field Definitions

This section defines the syntax and semantics of all commonly used HTTP/1.0 header fields. For general and entity header fields, both sender and recipient refer to either the client or the server, depending on who sends and who receives the message.

10.1 Allow

The Allow entity-header field lists the set of methods supported by the resource identified by the Request-URI. The purpose of this field is strictly to inform the recipient of valid methods associated with the resource. The Allow header field is not permitted in a request using the POST method, and thus should be ignored if it is received as part of a POST entity.

```
Allow           = "Allow" ":" 1#method
```

Example of use:

```
Allow: GET, HEAD
```

This field cannot prevent a client from trying other methods. However, the indications given by the Allow header field value should be followed. The actual set of allowed methods is defined by the origin server at the time of each request.

A proxy must not modify the Allow header field even if it does not understand all the methods specified, since the user agent may have other means of communicating with the origin server.

The Allow header field does not indicate what methods are implemented by the server.

10.2 Authorization

A user agent that wishes to authenticate itself with a server--usually, but not necessarily, after receiving a 401 response--may do so by including an Authorization request-header field with the request. The Authorization field value consists of credentials containing the authentication information of the user agent for the realm of the resource being requested.

```
Authorization  = "Authorization" ":" credentials
```

HTTP access authentication is described in Section 11. If a request is authenticated and a realm specified, the same credentials should be valid for all other requests within this realm.

Responses to requests containing an Authorization field are not cachable.

10.3 Content-Encoding

The Content-Encoding entity-header field is used as a modifier to the media-type. When present, its value indicates what additional content coding has been applied to the resource, and thus what decoding mechanism must be applied in order to obtain the media-type referenced by the Content-Type header field. The Content-Encoding is primarily used to allow a document to be compressed without losing the identity of its underlying media type.

```
Content-Encoding = "Content-Encoding" ":" content-coding
```

Content codings are defined in Section 3.5. An example of its use is

```
Content-Encoding: x-gzip
```

The Content-Encoding is a characteristic of the resource identified by the Request-URI. Typically, the resource is stored with this encoding and is only decoded before rendering or analogous usage.

10.4 Content-Length

The Content-Length entity-header field indicates the size of the Entity-Body, in decimal number of octets, sent to the recipient or, in the case of the HEAD method, the size of the Entity-Body that would have been sent had the request been a GET.

```
Content-Length = "Content-Length" ":" 1*DIGIT
```

An example is

```
Content-Length: 3495
```

Applications should use this field to indicate the size of the Entity-Body to be transferred, regardless of the media type of the entity. A valid Content-Length field value is required on all HTTP/1.0 request messages containing an entity body.

Any Content-Length greater than or equal to zero is a valid value. Section 7.2.2 describes how to determine the length of a response entity body if a Content-Length is not given.

Note: The meaning of this field is significantly different from the corresponding definition in MIME, where it is an optional field used within the "message/external-body" content-type. In HTTP, it should be used whenever the entity's length can be determined prior to being transferred.

10.5 Content-Type

The Content-Type entity-header field indicates the media type of the Entity-Body sent to the recipient or, in the case of the HEAD method, the media type that would have been sent had the request been a GET.

```
Content-Type   = "Content-Type" ":" media-type
```

Media types are defined in Section 3.6. An example of the field is

```
Content-Type: text/html
```

Further discussion of methods for identifying the media type of an entity is provided in Section 7.2.1.

10.6 Date

The Date general-header field represents the date and time at which the message was originated, having the same semantics as orig-date in RFC 822. The field value is an HTTP-date, as described in Section 3.3.

```
Date           = "Date" ":" HTTP-date
```

An example is

```
Date: Tue, 15 Nov 1994 08:12:31 GMT
```

If a message is received via direct connection with the user agent (in the case of requests) or the origin server (in the case of responses), then the date can be assumed to be the current date at the receiving end. However, since the date--as it is believed by the origin--is important for evaluating cached responses, origin servers should always include a Date header. Clients should only send a Date header field in messages that include an entity body, as in the case of the POST request, and even then it is optional. A received message which does not have a Date header field should be assigned one by the recipient if the message will be cached by that recipient or gatewayed via a protocol which requires a Date.

In theory, the date should represent the moment just before the entity is generated. In practice, the date can be generated at any time during the message origination without affecting its semantic value.

10.7 Expires

The Expires entity-header field gives the date/time after which the entity should be considered stale. This allows information providers to suggest the volatility of the resource, or a date after which the information may no longer be valid. Applications must not cache this entity beyond the date given. The presence of an Expires field does not imply that the original resource will change or cease to exist at, before, or after that time. However, information providers that know or even suspect that a resource will change by a certain date should include an Expires header with that date. The format is an absolute date and time as defined by HTTP-date in Section 3.3.

```
Expires        = "Expires" ":" HTTP-date
```

An example of its use is

```
Expires: Thu, 01 Dec 1994 16:00:00 GMT
```

If the date given is equal to or earlier than the value of the Date header, the recipient must not cache the enclosed entity. If a resource is dynamic by nature, as is the case with many data-producing processes, entities from that resource should be given an appropriate Expires value which reflects that dynamism.

The Expires field cannot be used to force a user agent to refresh its display or reload a resource; its semantics apply only to caching mechanisms, and such mechanisms need only check a resource's expiration status when a new request for that resource is initiated.

User agents often have history mechanisms, such as "Back" buttons and history lists, which can be used to redisplay an entity retrieved earlier in a session. By default, the Expires field does not apply to history mechanisms. If the entity is still in storage, a history mechanism should display it even if the entity has expired, unless the user has specifically configured the agent to refresh expired history documents.

Note: Applications are encouraged to be tolerant of bad or misinformed implementations of the Expires header. A value of zero (0) or an invalid date format should be considered equivalent to an "expires immediately." Although these values are not legitimate for HTTP/1.0, a robust implementation is always desirable.

10.8 From

The From request-header field, if given, should contain an Internet e-mail address for the human user who controls the requesting user agent. The address should be machine-usable, as defined by mailbox in RFC 822 (as updated by RFC 1123):

```
From            = "From" ":" mailbox
```

An example is:

```
From: webmaster@w3.org
```

This header field may be used for logging purposes and as a means for identifying the source of invalid or unwanted requests. It should not be used as an insecure form of access protection. The interpretation of this field is that the request is being performed on behalf of the person given, who accepts responsibility for the method performed. In particular, robot agents should include this header so that the person responsible for running the robot can be contacted if problems occur on the receiving end.

The Internet e-mail address in this field may be separate from the Internet host which issued the request. For example, when a request is passed through a proxy, the original issuer's address should be used.

10.9 If-Modified-Since

The If-Modified-Since request-header field is used with the GET method to make it conditional: if the requested resource has not been modified since the time specified in this field, a copy of the resource will not be returned from the server; instead, a 304 (not modified) response will be returned without any Entity-Body.

```
If-Modified-Since = "If-Modified-Since" ":" HTTP-date
```

An example of the field is:

```
If-Modified-Since: Sat, 29 Oct 1994 19:43:31 GMT
```

A conditional GET method requests that the identified resource be transferred only if it has been modified since the date given by the If-Modified-Since header. The algorithm for determining this includes the following cases:

a) If the request would normally result in anything other than a 200 (ok) status, or if the passed If-Modified-Since date is invalid, the response is exactly the same as for a normal GET. A date which is later than the server's current time is invalid.

b) If the resource has been modified since the If-Modified-Since date, the response is exactly the same as for a normal GET.

c) If the resource has not been modified since a valid If-Modified-Since date, the server shall return a 304 (not modified) response. The purpose of this feature is to allow efficient updates of cached information with a minimum amount of transaction overhead.

10.10 Last-Modified

The Last-Modified entity-header field indicates the date and time at which the sender believes the resource was last modified. The exact semantics of this field are defined in terms of how the recipient should interpret it: if the recipient has a copy of this resource which is older than the date given by the Last-Modified field, that copy should be considered stale.

```
Last-Modified  = "Last-Modified" ":" HTTP-date
```

An example of its use is

```
Last-Modified: Tue, 15 Nov 1994 12:45:26 GMT
```

The exact meaning of this header field depends on the implementation of the sender and the nature of the original resource. For files, it may be just the file system last-modified time. For entities with dynamically included parts, it may be the most recent of the set of last-modify times for its component parts. For database gateways, it may be the last-update timestamp of the record. For virtual objects, it may be the last time the internal state changed.

An origin server must not send a Last-Modified date which is later than the server's time of message origination. In such cases, where the resource's last modification would indicate some time in the future, the server must replace that date with the message origination date.

10.11 Location

The Location response-header field defines the exact location of the resource that was identified by the Request-URI. For 3xx responses, the location must indicate the server's preferred URL for automatic redirection to the resource. Only one absolute URL is allowed.

```
Location      = "Location" ":" absoluteURI
```

An example is

```
Location: http://www.w3.org/hypertext/WWW/NewLocation.html
```

10.12 Pragma

The Pragma general-header field is used to include implementation-specific directives that may apply to any recipient along the request/response chain. All pragma directives specify optional behavior from the viewpoint of the protocol; however, some systems may require that behavior be consistent with the directives.

```
Pragma           = "Pragma" ":" 1#pragma-directive
pragma-directive = "no-cache" | extension-pragma
extension-pragma = token [ "=" word ]
```

When the "no-cache" directive is present in a request message, an application should forward the request toward the origin server even if it has a cached copy of what is being requested. This allows a client to insist upon receiving an authoritative response to its request. It also allows a client to refresh a cached copy which is known to be corrupted or stale.

Pragma directives must be passed through by a proxy or gateway application, regardless of their significance to that application, since the directives may be applicable to all recipients along the request/response chain. It is not possible to specify a pragma for a specific recipient; however, any pragma directive not relevant to a recipient should be ignored by that recipient.

10.13 Referer

The Referer request-header field allows the client to specify, for the server's benefit, the address (URI) of the resource from which the Request-URI was obtained. This allows a server to generate lists of back-links to resources for interest, logging, optimized caching, etc. It also allows obsolete or mistyped links to be traced for maintenance. The Referer field must not be sent if the Request-URI was obtained from a source that does not have its own URI, such as input from the user keyboard.

```
Referer        = "Referer" ":" ( absoluteURI | relativeURI )
```

Example:

```
Referer: http://www.w3.org/hypertext/DataSources/Overview.html
```

If a partial URI is given, it should be interpreted relative to the Request-URI. The URI must not include a fragment.

10.14 Server

The Server response-header field contains information about the software used by the origin server to handle the request. The field can contain multiple product tokens (Section 3.7) and comments identifying the server and any significant subproducts. By convention, the product tokens are listed in order of their significance for identifying the application.

```
Server         = "Server" ":" 1*( product | comment )
```

Example:

```
Server: CERN/3.0 libwww/2.17
```

If the response is being forwarded through a proxy, the proxy application must not add its data to the product list.

10.15 User-Agent

The User-Agent request-header field contains information about the user agent originating the request. This is for statistical purposes, the tracing of protocol violations, and automated recognition of user agents for the sake of tailoring responses to avoid particular user agent limitations. Although it is not required, user agents should include this field with requests. The field can contain multiple product tokens (Section 3.7) and comments identifying the agent and any subproducts which form a significant part of the user agent. By convention, the product tokens are listed in order of their significance for identifying the application.

```
User-Agent     = "User-Agent" ":" 1*( product | comment )
```

Example:

```
User-Agent: CERN-LineMode/2.15 libwww/2.17b3
```

Note: Some current proxy applications append their product information to the list in the User-Agent field. This is not recommended, since it makes machine interpretation of these fields ambiguous.

10.16 WWW-Authenticate

The WWW-Authenticate response-header field must be included in 401 (unauthorized) response messages. The field value consists of at least one challenge that indicates the authentication scheme(s) and parameters applicable to the Request-URI.

```
WWW-Authenticate = "WWW-Authenticate" ":" 1#challenge
```

The HTTP access authentication process is described in Section 11. User agents must take special care in parsing the WWW-Authenticate field value if it contains more than one challenge, or if more than one WWW-Authenticate header field is provided, since the contents of a challenge may itself contain a comma-separated list of authentication parameters.

11. Access Authentication

HTTP provides a simple challenge-response authentication mechanism which may be used by a server to challenge a client request and by a client to provide authentication information. It uses an extensible, case-insensitive token to identify the authentication scheme, followed by a comma-separated list of attribute-value pairs which carry the parameters necessary for achieving authentication via that scheme.

```
auth-scheme    = token
auth-param     = token "=" quoted-string
```

The 401 (unauthorized) response message is used by an origin server to challenge the authorization of a user agent. This response must include a WWW-Authenticate header field containing at least one challenge applicable to the requested resource.

```
challenge       = auth-scheme 1*SP realm *( "," auth-param )

realm           = "realm" "=" realm-value
realm-value     = quoted-string
```

The realm attribute (case-insensitive) is required for all authentication schemes which issue a challenge. The realm value (case-sensitive), in combination with the canonical root URL of the server being accessed, defines the protection space. These realms allow the protected resources on a server to be partitioned into a set of protection spaces, each with its own authentication scheme and/or authorization database. The realm value is a string, generally assigned by the origin server, which may have additional semantics specific to the authentication scheme.

A user agent that wishes to authenticate itself with a server--usually, but not necessarily, after receiving a 401 response--may do so by including an Authorization header field with the request. The Authorization field value consists of credentials containing the authentication information of the user agent for the realm of the resource being requested.

```
credentials     = basic-credentials
                | ( auth-scheme #auth-param )
```

The domain over which credentials can be automatically applied by a user agent is determined by the protection space. If a prior request has been authorized, the same credentials may be reused for all other requests within that protection space for a period of time determined by the authentication scheme, parameters, and/or user preference. Unless otherwise defined by the authentication scheme, a single protection space cannot extend outside the scope of its server.

If the server does not wish to accept the credentials sent with a request, it should return a 403 (forbidden) response.

The HTTP protocol does not restrict applications to this simple challenge-response mechanism for access authentication. Additional mechanisms may be used, such as encryption at the transport level or via message encapsulation, and with additional header fields specifying authentication information. However, these additional mechanisms are not defined by this specification.

Proxies must be completely transparent regarding user agent authentication. That is, they must forward the WWW-Authenticate and Authorization headers untouched, and must not cache the response to a request containing Authorization. HTTP/1.0 does not provide a means for a client to be authenticated with a proxy.

11.1 Basic Authentication Scheme

The "basic" authentication scheme is based on the model that the user agent must authenticate itself with a user-ID and a password for each realm. The realm value should be considered an opaque string which can only be compared for equality with other realms on that server. The server will authorize the request only if it can validate the user-ID and password for the protection space of the Request-URI. There are no optional authentication parameters.

Upon receipt of an unauthorized request for a URI within the protection space, the server should respond with a challenge like the following:

```
WWW-Authenticate: Basic realm="WallyWorld"
```

where "WallyWorld" is the string assigned by the server to identify the protection space of the Request-URI.

To receive authorization, the client sends the user-ID and password, separated by a single colon (":") character, within a base64 encoded string in the credentials.

```
basic-credentials = "Basic" SP basic-cookie

basic-cookie      = <base64 [5] encoding of userid-password,
                     except not limited to 76 char/line>

userid-password   = [ token ] ":" *TEXT
```

If the user agent wishes to send the user-ID "Aladdin" and password "open sesame", it would use the following header field:

```
Authorization: Basic QWxhZGRpbjpvcGVuIHNlc2FtZQ==
```

The basic authentication scheme is a non-secure method of filtering unauthorized access to resources on an HTTP server. It is based on the assumption that the connection between the client and the server can be regarded as a trusted carrier. As this is not generally true on an open network, the basic authentication scheme should be used accordingly. In spite of this, clients should implement the scheme in order to communicate with servers that use it.

Appendix B – DNS – Implementation and Specification

This specification of the DNS protocol is based on RFC 1035 "Domain Names – Implementation and Specification" from the Network Working Group, 1987. Certain sections have been shortened or omitted, and readers are encouraged to consult the full document at http://www.ietf.org/rfc/rfc1035.txt

Status of this Memo

This RFC describes the details of the domain system and protocol, and assumes that the reader is familiar with the concepts discussed in a companion RFC, "Domain Names - Concepts and Facilities" [RFC-1034].

The domain system is a mixture of functions and data types which are an official protocol and functions and data types which are still experimental. Since the domain system is intentionally extensible, new data types and experimental behavior should always be expected in parts of the system beyond the official protocol. The official protocol parts include standard queries, responses and the Internet class RR data formats (e.g., host addresses). Since the previous RFC set, several definitions have changed, so some previous definitions are obsolete.

Experimental or obsolete features are clearly marked in these RFCs, and such information should be used with caution.

The reader is especially cautioned not to depend on the values which appear in examples to be current or complete, since their purpose is primarily pedagogical. Distribution of this memo is unlimited.

Table of Contents

2. Introduction

2.1 Overview

The goal of domain names is to provide a mechanism for naming resources in such a way that the names are usable in different hosts, networks, protocol families, internets, and administrative organizations.

From the user's point of view, domain names are useful as arguments to a local agent, called a resolver, which retrieves information associated with the domain name. Thus a user might ask for the host address or mail information associated with a particular domain name. To enable the user to request a particular type of information, an appropriate query type is passed to the resolver with the domain name. To the user, the domain tree is a single information space; the resolver is responsible for hiding the distribution of data among name servers from the user.

From the resolver's point of view, the database that makes up the domain space is distributed among various name servers. Different parts of the domain space are stored in different name servers, although a particular data item will be stored redundantly in two or more name servers. The resolver starts with knowledge of at least one name server. When the resolver processes a user query it asks a known name server for the information; in return, the resolver either receives the desired information or a referral to another name server. Using these referrals, resolvers learn the identities and contents of other name servers. Resolvers are responsible for dealing with the distribution of the domain space and dealing with the effects of name server failure by consulting redundant databases in other servers.

Name servers manage two kinds of data. The first kind of data held in sets called zones; each zone is the complete database for a particular "pruned" subtree of the domain space. This data is called authoritative. A name server periodically checks to make sure that its zones are up to date, and if not, obtains a new copy of updated zones from master files stored locally or in another name server. The second kind of data is cached data which was acquired by a local resolver. This data may be incomplete, but improves the performance of the retrieval process when non-local data is repeatedly accessed. Cached data is eventually discarded by a timeout mechanism.

This functional structure isolates the problems of user interface, failure recovery, and distribution in the resolvers and isolates the database update and refresh problems in the name servers.

2.2 Common configurations

A host can participate in the domain name system in a number of ways, depending on whether the host runs programs that retrieve information from the domain system, name servers that answer queries from other hosts, or various combinations of both functions. The simplest, and perhaps most typical, configuration is shown below:

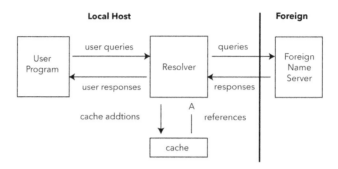

User programs interact with the domain name space through resolvers; the format of user queries and user responses is specific to the host and its operating system. User queries will typically be operating system calls, and the resolver and its cache will be part of the host operating system. Less capable hosts may choose to implement the resolver as a subroutine to be linked in with every program that needs its services. Resolvers answer user queries with information they acquire via queries to foreign name servers and the local cache.

Note that the resolver may have to make several queries to several different foreign name servers to answer a particular user query, and hence the resolution of a user query may involve several network accesses and an arbitrary amount of time. The queries to foreign name servers and the corresponding responses have a standard format described in this memo, and may be datagrams.

Depending on its capabilities, a name server could be a stand alone program on a dedicated machine or a process or processes on a large timeshared host. A simple configuration might be:

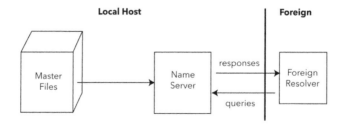

Here a primary name server acquires information about one or more zones by reading master files from its local file system, and answers queries about those zones that arrive from foreign resolvers.

The DNS requires that all zones be redundantly supported by more than one name server. Designated secondary servers can acquire zones and check for updates from the primary server using the zone transfer protocol of the DNS. This configuration is shown below:

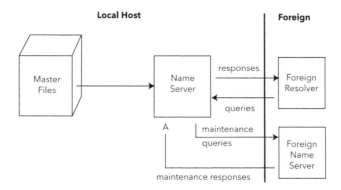

In this configuration, the name server periodically establishes a virtual circuit to a foreign name server to acquire a copy of a zone or to check that an existing copy has not changed. The messages sent for these maintenance activities follow the same form as queries and responses, but the message sequences are somewhat different.

The information flow in a host that supports all aspects of the domain name system is shown below:

186

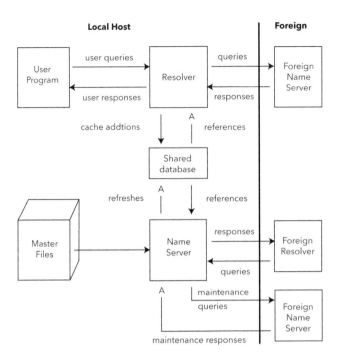

The shared database holds domain space data for the local name server and resolver. The contents of the shared database will typically be a mixture of authoritative data maintained by the periodic refresh operations of the name server and cached data from previous resolver requests. The structure of the domain data and the necessity for synchronization between name servers and resolvers imply the general characteristics of this database, but the actual format is up to the local implementor.

Information flow can also be tailored so that a group of hosts act together to optimize activities. Sometimes this is done to offload less capable hosts so that they do not have to implement a full resolver. This can be appropriate for PCs or hosts which want to minimize the amount of new network code which is required. This scheme can also allow a group of hosts can share a small number of caches rather than maintaining a large number of separate caches, on the premise that the centralized caches will have a higher hit ratio. In either case, resolvers are replaced with stub resolvers which act as front ends to resolvers located in a recursive server in one or more name servers known to perform that service.

In any case, note that domain components are always replicated for reliability whenever possible.

2.3 Conventions

The domain system has several conventions dealing with low-level, but fundamental, issues. While the implementor is free to violate theseconventions WITHIN HIS OWN SYSTEM, he must observe these conventions in ALL behavior observed from other hosts.

2.3.1 Preferred name syntax

The DNS specifications attempt to be as general as possible in the rules for constructing domain names. The idea is that the name of any existing object can be expressed as a domain name with minimal changes. However, when assigning a domain name for an object, the prudent user will select a name which satisfies both the rules of the domain system and any existing rules for the object, whether these rules are published or implied by existing programs.

For example, when naming a mail domain, the user should satisfy both the rules of this memo and those in RFC-822. When creating a new host name, the old rules for HOSTS.TXT should be followed. This avoids problems when old software is converted to use domain names.

The following syntax will result in fewer problems with many applications that use domain names (e.g., mail, TELNET).

```
<domain> ::= <subdomain> | " "
<subdomain> ::= <label> | <subdomain> "." <label>
<label> ::= <letter> [ [ <ldh-str> ] <let-dig> ]
<ldh-str> ::= <let-dig-hyp> | <let-dig-hyp> <ldh-str>
<let-dig-hyp> ::= <let-dig> | "-"
<let-dig> ::= <letter> | <digit>
<letter> ::= any one of the 52 alphabetic characters A
             through Z in upper case and a through z in
             lower case
<digit> ::= any one of the ten digits 0 through 9
```

Note that while upper and lower case letters are allowed in domain names, no significance is attached to the case. That is, two names with the same spelling but different case are to be treated as if identical.

The labels must follow the rules for ARPANET host names. They must start with a letter, end with a letter or digit, and have as interior characters only letters, digits, and hyphen. There are also some restrictions on the length. Labels must be 63 characters or less.

2.3.2 Data Transmission Order

The order of transmission of the header and data described in this document is resolved to the octet level. Whenever a diagram shows a group of octets, the order of transmission of those octets is the normal order in which they are read in English. For example, in the following diagram, the octets are transmitted in the order they are numbered.

```
    0                   1
    0 1 2 3 4 5 6 7 8 9 0 1 2 3 4 5
   +-+-+-+-+-+-+-+-+-+-+-+-+-+-+-+-+
   |       1       |       2       |
   +-+-+-+-+-+-+-+-+-+-+-+-+-+-+-+-+
   |       3       |       4       |
   +-+-+-+-+-+-+-+-+-+-+-+-+-+-+-+-+
   |       5       |       6       |
   +-+-+-+-+-+-+-+-+-+-+-+-+-+-+-+-+
```

Whenever an octet represents a numeric quantity, the left most bit in the diagram is the high order or most significant bit. That is, the bit labeled 0 is the most significant bit. For example, the following diagram represents the value 170 (decimal).

```
    0 1 2 3 4 5 6 7
   +-+-+-+-+-+-+-+-+
   |1 0 1 0 1 0 1 0|
   +-+-+-+-+-+-+-+-+
```

Similarly, whenever a multi-octet field represents a numeric quantity the left most bit of the whole field is the most significant bit. When a multi-octet quantity is transmitted the most significant octet is transmitted first.

2.3.3 Character Case

For all parts of the DNS that are part of the official protocol, all comparisons between character strings (e.g., labels, domain names, etc.) are done in a case-insensitive manner. At present, this rule is in force throughout the domain system without exception. However, future additions beyond current usage may need to use the full binary octet capabilities in names, so attempts to store domain names in 7-bit ASCII or use of special bytes to terminate labels, etc., should be avoided.

When data enters the domain system, its original case should be preserved whenever possible. In certain circumstances this cannot be done. For example, if two RRs are stored in a database, one at x.y and one at X.Y, they are actually stored at the same place in the database, and hence only one casing would be preserved. The basic rule is that case can be discarded only when data is used to define structure in a database, and two names are identical when compared in a case insensitive manner.

Loss of case sensitive data must be minimized. Thus while data for x.y and X.Y may both be stored under a single location x.y or X.Y, data for a.x and B.X would never be stored under A.x, A.X, b.x, or b.X. In general, this preserves the case of the first label of a domain name, but forces standardization of interior node labels.

Systems administrators who enter data into the domain database should take care to represent the data they supply to the domain system in a case-consistent manner if their system is case-sensitive. The data distribution system in the domain system will ensure that consistent representations are preserved.

2.3.4 Size limits

Various objects and parameters in the DNS have size limits. They are listed below. Some could be easily changed, others are more fundamental.

labels
    63 octets or less

names
    255 octets or less

TTL
    positive values of a signed 32 bit number.

UDP messages
    512 octets or less

3. Domain Name Space and RR Definitions

3.1 Name space definitions

Domain names in messages are expressed in terms of a sequence of labels. Each label is represented as a one octet length field followed by that number of octets. Since every domain name ends with the null label of the root, a domain name is terminated by a length byte of zero. The high order two bits of every length octet must be zero, and the remaining six bits of the length field limit the label to 63 octets or less.

To simplify implementations, the total length of a domain name (i.e., label octets and label length octets) is restricted to 255 octets or less.

Although labels can contain any 8 bit values in octets that make up a label, it is strongly recommended that labels follow the preferred syntax described elsewhere in this memo, which is compatible with existing host naming conventions. Name servers and resolvers must compare labels in a case-insensitive manner (i.e., A=a), assuming ASCII with zero parity. Non-alphabetic codes must match exactly.

3.2 RR Definitions

3.2.1 Format

All RRs have the same top level format shown below:

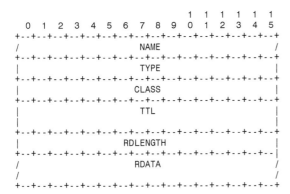

```
                                    1  1  1  1  1  1
    0  1  2  3  4  5  6  7  8  9  0  1  2  3  4  5
  +--+--+--+--+--+--+--+--+--+--+--+--+--+--+--+--+
  /                     NAME                      /
  +--+--+--+--+--+--+--+--+--+--+--+--+--+--+--+--+
  |                     TYPE                      |
  +--+--+--+--+--+--+--+--+--+--+--+--+--+--+--+--+
  |                     CLASS                     |
  +--+--+--+--+--+--+--+--+--+--+--+--+--+--+--+--+
  |                     TTL                       |
  |                                               |
  +--+--+--+--+--+--+--+--+--+--+--+--+--+--+--+--+
  |                   RDLENGTH                    |
  +--+--+--+--+--+--+--+--+--+--+--+--+--+--+--+--|
  /                     RDATA                     /
  /                                               /
  +--+--+--+--+--+--+--+--+--+--+--+--+--+--+--+--+
```

where:

NAME

an owner name, i.e., the name of the node to which this resource record pertains.

TYPE

two octets containing one of the RR TYPE codes.

CLASS

two octets containing one of the RR CLASS codes.

TTL

a 32 bit signed integer that specifies the time interval that the resource record may be cached before the source of the information should again be consulted. Zero values are interpreted to mean that the RR can only be used for the transaction in progress, and should not be cached. For example, SOA records are always distributed with a zero TTL to prohibit caching. Zero values can also be used for extremely volatile data.

RDLENGTH

an unsigned 16 bit integer that specifies the length in octets of the RDATA field.

RDATA

a variable length string of octets that describes the resource. The format of this information varies according to the TYPE and CLASS of the resource record.

3.2.2 TYPE values

TYPE fields are used in resource records. Note that these types are a subset of QTYPEs.

TYPE	Value	Meaning
A	1	a host address
NS	2	an authoritative name server
MD	3	a mail destination (Obsolete - use MX)
MF	4	a mail forwarder (Obsolete - use MX)
CNAME	5	the canonical name for an alias
SOA	6	marks the start of a zone of authority
MB	7	a mailbox domain name (EXPERIMENTAL)
MG	8	a mail group member (EXPERIMENTAL)
MR	9	a mail rename domain name (EXPERIMENTAL)
NULL	10	a null RR (EXPERIMENTAL)
WKS	11	a well known service description
PTR	12	a domain name pointer
HINFO	13	host information
MINFO	14	mailbox or mail list information
MX	15	mail exchange
TXT	16	text strings

3.2.3 QTYPE values

QTYPE fields appear in the question part of a query. QTYPES are a superset of TYPEs, hence all TYPEs are valid QTYPEs. In addition, the following QTYPEs are defined:

QTYPE	Value	Meaning
AXFR	252	A request for a transfer of an entire zone
MAILB	253	A request for mailbox-related records (MB, MG or MR)
MAILA	254	A request for mail agent RRs (Obsolete - see MX)

QTYPE	Value	Meaning
*	255	A request for all records

3.2.4 CLASS values

CLASS fields appear in resource records. The following CLASS mnemonics and values are defined:

CLASS	Value	Meaning
IN	1	the Internet
CS	2	the CSNET class (Obsolete - used only for examples in some obsolete RFCs)
CH	3	the CHAOS class
HS	4	Hesiod [Dyer 87]

3.2.5 QCLASS values

QCLASS fields appear in the question section of a query. QCLASS values are a superset of CLASS values; every CLASS is a valid QCLASS. In addition to CLASS values, the following QCLASSes are defined:

QCLASS	Value	Meaning
*	255	any class

3.3 Standard RRs

The following RR definitions are expected to occur, at least potentially, in all classes. In particular, NS, SOA, CNAME, and PTR will be used in all classes, and have the same format in all classes. Because their RDATA format is known, all domain names in the RDATA section of these RRs may be compressed.

<domain-name> is a domain name represented as a series of labels, and terminated by a label with zero length. <character-string> is a single length octet followed by that number of characters. <character-string> is treated as binary information, and can be up to 256 characters in length (including the length octet).

3.3.1 CNAME RDATA format

CNAME

A <domain-name> which specifies the canonical or primary name for the owner. The owner name is an alias.

CNAME RRs cause no additional section processing, but name servers may choose to restart the query at the canonical name in certain cases. See the description of name server logic in [RFC-1034] for details.

3.3.2 HINFO RDATA format

CPU

A <character-string> which specifies the CPU type.

OS

A <character-string> which specifies the operating system type.

Standard values for CPU and OS can be found in [RFC-1010].

HINFO records are used to acquire general information about a host. The main use is for protocols such as FTP that can use special procedures when talking between machines or operating systems of the same type.

3.3.3 MB RDATA format (EXPERIMENTAL)

MADNAME

A <domain-name> which specifies a host which has the specified mailbox.

MB records cause additional section processing which looks up an A type RRs corresponding to MADNAME.

3.3.4 MD RDATA format (Obsolete)

MADNAME

A <domain-name> which specifies a host which has a mail agent for the domain which should be able to deliver mail for the domain.

MD records cause additional section processing which looks up an A type record corresponding to MADNAME.

MD is obsolete. See the definition of MX and [RFC-974] for details of the new scheme. The recommended policy for dealing with MD RRs found in a master file is to reject them, or to convert them to MX RRs with a preference of 0.

3.3.5 MF RDATA format (Obsolete)

MADNAME

A <domain-name> which specifies a host which has a mail agent for the domain which will accept mail for forwarding to the domain.

MF records cause additional section processing which looks up an A type record corresponding to MADNAME.

MF is obsolete. See the definition of MX and [RFC-974] for details of the new scheme. The recommended policy for dealing with MD RRs found in a master file is to reject them, or to convert them to MX RRs with a preference of 10.

3.3.6 MG RDATA format (EXPERIMENTAL)

MGMNAME

A <domain-name> which specifies a mailbox which is a member of the mail group specified by the domain name.

MG records cause no additional section processing.

3.3.7 MINFO RDATA format (EXPERIMENTAL)

RMAILBX

A <domain-name> which specifies a mailbox which is responsible for the mailing list or mailbox. If this domain name names the root, the owner of the MINFO RR is responsible for itself. Note that many existing mailing lists use a mailbox X-request for the RMAILBX field of mailing list X, e.g., Msgroup-request for Msgroup. This field provides a more general mechanism.

EMAILBX

A <domain-name> which specifies a mailbox which is to receive error messages related to the mailing list or mailbox specified by the owner of the MINFO RR (similar to the ERRORS-TO: field which has been proposed). If this domain name names the root, errors should be returned to the sender of the message.

MINFO records cause no additional section processing. Although these records can be associated with a simple mailbox, they are usually used with a mailing list.

3.3.8 MR RDATA format (EXPERIMENTAL)

NEWNAME

A <domain-name> which specifies a mailbox which is the proper rename of the specified mailbox.

MR records cause no additional section processing. The main use for MR is as a forwarding entry for a user who has moved to a different mailbox.

3.3.9 MX RDATA format

PREFERENCE

A 16 bit integer which specifies the preference given to this RR among others at the same owner. Lower values are preferred.

EXCHANGE

A <domain-name> which specifies a host willing to act as a mail exchange for the owner name.

MX records cause type A additional section processing for the host specified by EXCHANGE. The use of MX RRs is explained in detail in [RFC-974].

3.3.10 NULL RDATA format (EXPERIMENTAL)

Anything at all may be in the RDATA field so long as it is 65535 octets or less.

NULL records cause no additional section processing. NULL RRs are not allowed in master files. NULLs are used as placeholders in some experimental extensions of the DNS.

3.3.11 NS RDATA format

NSDNAME

A <domain-name> which specifies a host which should be authoritative for the specified class and domain.

NS records cause both the usual additional section processing to locate a type A record, and, when used in a referral, a special search of the zone in which they reside for glue information.

The NS RR states that the named host should be expected to have a zone starting at owner name of the specified class. Note that the class may not indicate the protocol family which should be used to communicate with the host, although it is typically a strong hint. For example, hosts which are name servers for either Internet (IN) or Hesiod (HS) class information are normally queried using IN class protocols.

3.3.12 PTR RDATA format

PTRDNAME

A <domain-name> which points to some location in the domain name space.

PTR records cause no additional section processing. These RRs are used in special domains to point to some other location in the domain space.These records are simple data, and don't imply any special processing similar to that performed by CNAME, which identifies aliases. See the description of the IN-ADDR.ARPA domain for an example.

3.3.13 SOA RDATA format

MNAME

The <domain-name> of the name server that was the original or primary source of data for this zone.

RNAME

A <domain-name> which specifies the mailbox of the person responsible for this zone.

SERIAL

The unsigned 32 bit version number of the original copy of the zone. Zone transfers preserve this value. This value wraps and should be compared using sequence space arithmetic.

REFRESH

A 32 bit time interval before the zone should be refreshed.

RETRY

A 32 bit time interval that should elapse before a failed refresh should be retried.

EXPIRE

A 32 bit time value that specifies the upper limit on the time interval that can elapse before the zone is no longer authoritative.

MINIMUM

The unsigned 32 bit minimum TTL field that should be exported with any RR from this zone.

SOA records cause no additional section processing.

All times are in units of seconds.

Most of these fields are pertinent only for name server maintenance operations. However, MINIMUM is used in all query operations that retrieve RRs from a zone. Whenever a RR is sent in a response to a query, the TTL field is set to the maximum of the TTL field from the RR and the

MINIMUM field in the appropriate SOA. Thus MINIMUM is a lower bound on the TTL field for all RRs in a zone. Note that this use of MINIMUM should occur when the RRs are copied into the response and not when the zone is loaded from a master file or via a zone transfer. The reason for this provison is to allow future dynamic update facilities to change the SOA RR with known semantics.

3.3.14 TXT RDATA format

TXT-DATA
    One or more <character-string>s.

TXT RRs are used to hold descriptive text. The semantics of the text depends on the domain where it is found.

3.4 Internet specific RRs

3.4.1 A RDATA format

ADDRESS
    A 32 bit Internet address.

Hosts that have multiple Internet addresses will have multiple A records.

A records cause no additional section processing. The RDATA section of an A line in a master file is an Internet address expressed as four decimal numbers separated by dots without any imbedded spaces (e.g., "10.2.0.52" or "192.0.5.6").

3.4.2 WKS RDATA format

ADDRESS
    A 32-bit Internet address

PROTOCOL
    An 8-bit IP protocol number

<BIT MAP>
    A variable length bit map. The bit map must be a multiple of 8 bits long.

The WKS record is used to describe the well known services supported by a particular protocol on a particular internet address. The PROTOCOL field specifies an IP protocol number, and the bit map has one bit per port of the specified protocol. The first bit corresponds to port 0, the second to port 1, etc. If the bit map does not include a bit for a protocol of interest, that bit is assumed zero. The appropriate values and mnemonics for ports and protocols are specified in [RFC-1010].

For example, if PROTOCOL=TCP (6), the 26th bit corresponds to TCP port 25 (SMTP). If this bit is set, a SMTP server should be listening on TCP port 25; if zero, SMTP service is not supported on the specified address.

The purpose of WKS RRs is to provide availability information for servers for TCP and UDP. If a server supports both TCP and UDP, or has multiple Internet addresses, then multiple WKS RRs are used.

WKS RRs cause no additional section processing.

In master files, both ports and protocols are expressed using mnemonics or decimal numbers.

3.5 IN-ADDR.ARPA domain

The Internet uses a special domain to support gateway location and Internet address to host mapping. Other classes may employ a similar strategy in other domains. The intent of this domain is to provide a guaranteed method to perform host address to host name mapping, and to facilitate queries to locate all gateways on a particular network in the Internet.

Note that both of these services are similar to functions that could be performed by inverse queries; the difference is that this part of the domain name space is structured according to address, and hence can guarantee that the appropriate data can be located without an exhaustive search of the domain space.

The domain begins at IN-ADDR.ARPA and has a substructure which follows the Internet addressing structure.

Domain names in the IN-ADDR.ARPA domain are defined to have up to four labels in addition to the IN-ADDR.ARPA suffix. Each label represents one octet of an Internet address, and is expressed as a character string for a decimal value in the range 0-255 (with leading zeros omitted except in the case of a zero octet which is represented by a single zero).

Host addresses are represented by domain names that have all four labels specified. Thus data for Internet address 10.2.0.52 is located at domain name 52.0.2.10.IN-ADDR.ARPA. The reversal, though awkward to read, allows zones to be delegated which are exactly one network of address space. For example, 10.IN-ADDR.ARPA can be a zone containing data for the ARPANET, while

26.IN-ADDR.ARPA can be a separate zone for MILNET. Address nodes are used to hold pointers to primary host names in the normal domain space.

Network numbers correspond to some non-terminal nodes at various depths in the IN-ADDR.ARPA domain, since Internet network numbers are either 1, 2, or 3 octets. Network nodes are used to hold pointers to the primary host names of gateways attached to that network. Since a gateway is, by definition, on more than one network, it will typically have two or more network nodes which point at it. Gateways will also have host level pointers at their fully qualified addresses.

Both the gateway pointers at network nodes and the normal host pointers at full address nodes use the PTR RR to point back to the primary domain names of the corresponding hosts.

For example, the IN-ADDR.ARPA domain will contain information about the ISI gateway between net 10 and 26, an MIT gateway from net 10 to MIT's net 18, and hosts A.ISI.EDU and MULTICS.MIT.EDU. Assuming that ISI gateway has addresses 10.2.0.22 and 26.0.0.103, and a name MILNET-GW.ISI.EDU, and the MIT gateway has addresses 10.0.0.77 and 18.10.0.4 and a name GW.LCS.MIT.EDU, the domain database would contain:

```
10.IN-ADDR.ARPA.          PTR MILNET-GW.ISI.EDU.
10.IN-ADDR.ARPA.          PTR GW.LCS.MIT.EDU.
18.IN-ADDR.ARPA.          PTR GW.LCS.MIT.EDU.
26.IN-ADDR.ARPA.          PTR MILNET-GW.ISI.EDU.
22.0.2.10.IN-ADDR.ARPA.   PTR MILNET-GW.ISI.EDU.
103.0.0.26.IN-ADDR.ARPA.  PTR MILNET-GW.ISI.EDU.
77.0.0.10.IN-ADDR.ARPA.   PTR GW.LCS.MIT.EDU.
4.0.10.18.IN-ADDR.ARPA.   PTR GW.LCS.MIT.EDU.
103.0.3.26.IN-ADDR.ARPA.  PTR A.ISI.EDU.
6.0.0.10.IN-ADDR.ARPA.    PTR MULTICS.MIT.EDU.
```

Thus a program which wanted to locate gateways on net 10 would originate a query of the form QTYPE=PTR, QCLASS=IN, QNAME=10.IN-ADDR.ARPA. It would receive two RRs in response:

```
10.IN-ADDR.ARPA.          PTR MILNET-GW.ISI.EDU.
10.IN-ADDR.ARPA.          PTR GW.LCS.MIT.EDU.
```

The program could then originate QTYPE=A, QCLASS=IN queries for MILNET-GW.ISI.EDU. and GW.LCS.MIT.EDU. to discover the Internet addresses of these gateways.

A resolver which wanted to find the host name corresponding to Internet host address 10.0.0.6 would pursue a query of the form QTYPE=PTR, QCLASS=IN, QNAME=6.0.0.10.IN-ADDR.ARPA, and would receive:

```
6.0.0.10.IN-ADDR.ARPA.    PTR MULTICS.MIT.EDU.
```

Several cautions apply to the use of these services:

- Since the IN-ADDR.ARPA special domain and the normal domain for a particular host or gateway will be in different zones, the possibility exists that that the data may be inconsistent.

- Gateways will often have two names in separate domains, only one of which can be primary.

- Systems that use the domain database to initialize their routing tables must start with enough gateway information to guarantee that they can access the appropriate name server.

- The gateway data only reflects the existence of a gateway in a manner equivalent to the current HOSTS.TXT file. It doesn't replace the dynamic availability information from GGP or EGP.

4. Messages

4.1 Format

All communications inside of the domain protocol are carried in a single format called a message. The top level format of message is divided into 5 sections (some of which are empty in certain cases) shown below:

```
+---------------------+
|       Header        |
+---------------------+
|      Question       |
+---------------------+
|       Answer        |
+---------------------+
|      Authority      |
+---------------------+
|      Additional     |
+---------------------+
```

The header section is always present. The header includes fields that specify which of the remaining sections are present, and also specify whether the message is a query or a response, a standard query or some other opcode, etc.

The names of the sections after the header are derived from their use in standard queries. The question section contains fields that describe a question to a name server. These fields are a query type (QTYPE), a query class (QCLASS), and a query domain name (QNAME). The last three sections have the same format: a possibly empty list of concatenated resource records (RRs). The answer section contains RRs that answer the question; the authority section contains RRs that point toward an authoritative name server; the additional records section contains RRs which relate to the query, but are not strictly answers for the question.

4.1.1 Header section format

The header contains the following fields:

```
                                1 1 1 1  1  1
    0  1  2  3  4  5  6  7  8  9 0 1 2 3  4  5
  +--+--+--+--+--+--+--+--+--+--+--+--+--+--+--+--+
  |                      ID                       |
  +--+--+--+--+--+--+--+--+--+--+--+--+--+--+--+--+
  |QR|   Opcode  |AA|TC|RD|RA|   Z    |   RCODE   |
  +--+--+--+--+--+--+--+--+--+--+--+--+--+--+--+--+
  |                    QDCOUNT                     |
  +--+--+--+--+--+--+--+--+--+--+--+--+--+--+--+--+
  |                    ANCOUNT                     |
  +--+--+--+--+--+--+--+--+--+--+--+--+--+--+--+--+
  |                    NSCOUNT                     |
  +--+--+--+--+--+--+--+--+--+--+--+--+--+--+--+--+
  |                    ARCOUNT                     |
  +--+--+--+--+--+--+--+--+--+--+--+--+--+--+--+--+
```

where:

ID

A 16 bit identifier assigned by the program that generates any kind of query. This identifier is copied the corresponding reply and can be used by the requester to match up replies to outstanding queries.

QR

A one bit field that specifies whether this message is a query (0), or a response (1).

OPCODE

A four bit field that specifies kind of query in this message. This value is set by the originator of a query and copied into the response. The values are:

0 – a standard query (QUERY)

1 – an inverse query (IQUERY)

2 – a server status request (STATUS)

3-15 – reserved for future use

AA

Authoritative Answer - this bit is valid in responses, and specifies that the responding name server is an authority for the domain name in question section.

Note that the contents of the answer section may have multiple owner names because of aliases. The AA bit corresponds to the name which matches the query name, or the first owner name in the answer section.

TC

TrunCation - specifies that this message was truncated due to length greater than that permitted on the transmission channel.

RD

Recursion Desired - this bit may be set in a query and is copied into the response. If RD is set, it directs the name server to pursue the query recursively. Recursive query support is optional.

RA

Recursion Available - this be is set or cleared in a response, and denotes whether recursive query support is available in the name server.

Z

Reserved for future use. Must be zero in all queries and responses.

RCODE

Response code - this 4 bit field is set as part of responses. The values have the following interpretation:

0 – No error condition

1 – Format error - The name server was unable to interpret the query.

2 – Server failure - The name server was unable to process this query due to a problem with the name server.

3 – Name Error - Meaningful only for responses from an authoritative name server, this code signifies that the domain name referenced in the query does not exist.

4 – Not Implemented - The name server does not support the requested kind of query.

5 – Refused - The name server refuses to perform the specified operation for policy reasons. For example, a name server may not wish to provide the information to the particular requester, or a name server may not wish to perform a particular operation (e.g., zone transfer) for particular data.

6-15 – Reserved for future use.

QDCOUNT

an unsigned 16 bit integer specifying the number of entries in the question section.

ANCOUNT

an unsigned 16 bit integer specifying the number of resource records in the answer section.

NSCOUNT

an unsigned 16 bit integer specifying the number of name server resource records in the authority records section.

ARCOUNT

an unsigned 16 bit integer specifying the number of resource records in the additional records section.

4.1.2 Question section format

The question section is used to carry the "question" in most queries, i.e., the parameters that define what is being asked. The section contains QDCOUNT (usually 1) entries, each of the following format:

```
                                    1  1  1  1  1  1
    0  1  2  3  4  5  6  7  8  9  0  1  2  3  4  5
  +--+--+--+--+--+--+--+--+--+--+--+--+--+--+--+--+
  /                     QNAME                     /
  +--+--+--+--+--+--+--+--+--+--+--+--+--+--+--+--+
  |                     QTYPE                     |
  +--+--+--+--+--+--+--+--+--+--+--+--+--+--+--+--+
  |                     QCLASS                    |
  +--+--+--+--+--+--+--+--+--+--+--+--+--+--+--+--+
```

where:

QNAME

a domain name represented as a sequence of labels, where each label consists of a length octet followed by that number of octets. The domain name terminates with the zero length octet for the null label of the root. Note that this field may be an odd number of octets; no padding is used.

QTYPE

a two octet code which specifies the type of the query. The values for this field include all codes valid for a TYPE field, together with some more general codes which can match more than one type of RR.

QCLASS

a two octet code that specifies the class of the query. For example, the QCLASS field is IN for the Internet.

4.1.3 Resource record format

The answer, authority, and additional sections all share the same format: a variable number of resource records, where the number of records is specified in the corresponding count field in the header. Each resource record has the following format:

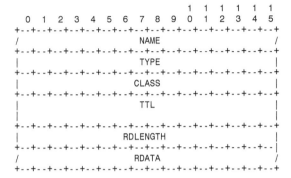

where:

NAME
   a domain name to which this resource record pertains.

TYPE
   two octets containing one of the RR type codes. This field specifies the meaning of the data in the RDATA field.

CLASS
   two octets which specify the class of the data in the RDATA field.

TTL
   a 32 bit unsigned integer that specifies the time interval (in seconds) that the resource record may be cached before it should be discarded. Zero values are interpreted to mean that the RR can only be used for the transaction in progress, and should not be cached.

RDLENGTH
   an unsigned 16 bit integer that specifies the length in octets of the RDATA field.

RDATA
   a variable length string of octets that describes the resource. The format of this information varies according to the TYPE and CLASS of the resource record. For example, the if the TYPE is A and the CLASS is IN, the RDATA field is a 4 octet ARPA Internet address.

4.1.4 Message compression

In order to reduce the size of messages, the domain system utilizes a compression scheme which eliminates the repetition of domain names in a message. In this scheme, an entire domain name or a list of labels at the end of a domain name is replaced with a pointer to a prior occurance of the same name.

The pointer takes the form of a two octet sequence:

```
+--+--+--+--+--+--+--+--+--+--+--+--+--+--+--+--+
| 1  1|                OFFSET                    |
+--+--+--+--+--+--+--+--+--+--+--+--+--+--+--+--+
```

The first two bits are ones. This allows a pointer to be distinguished from a label, since the label must begin with two zero bits because labels are restricted to 63 octets or less. (The 10 and 01 combinations are reserved for future use.) The OFFSET field specifies an offset from the start of the message (i.e., the first octet of the ID field in the domain header). A zero offset specifies the first byte of the ID field, etc.

The compression scheme allows a domain name in a message to be represented as either:

- a sequence of labels ending in a zero octet
- a pointer
- a sequence of labels ending with a pointer

Pointers can only be used for occurances of a domain name where the format is not class specific. If this were not the case, a name server or resolver would be required to know the format of all RRs it handled. As yet, there are no such cases, but they may occur in future RDATA formats.

If a domain name is contained in a part of the message subject to a length field (such as the RDATA section of an RR), and compression is used, the length of the compressed name is used in the length calculation, rather than the length of the expanded name.

Programs are free to avoid using pointers in messages they generate, although this will reduce datagram capacity, and may cause truncation. However all programs are required to understand arriving messages that contain pointers.

For example, a datagram might need to use the domain names F.ISI.ARPA, FOO.F.ISI.ARPA, ARPA, and the root. Ignoring the other fields of the message, these domain names might be represented as:

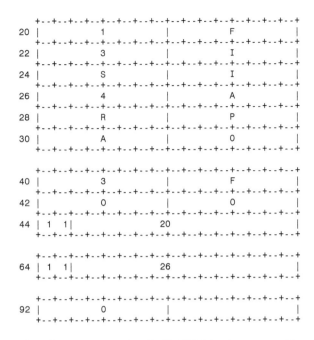

```
       +--+--+--+--+--+--+--+--+--+--+--+--+--+--+--+--+
    20 |              1            |              F            |
       +--+--+--+--+--+--+--+--+--+--+--+--+--+--+--+--+
    22 |              3            |              I            |
       +--+--+--+--+--+--+--+--+--+--+--+--+--+--+--+--+
    24 |              S            |              I            |
       +--+--+--+--+--+--+--+--+--+--+--+--+--+--+--+--+
    26 |              4            |              A            |
       +--+--+--+--+--+--+--+--+--+--+--+--+--+--+--+--+
    28 |              R            |              P            |
       +--+--+--+--+--+--+--+--+--+--+--+--+--+--+--+--+
    30 |              A            |              O            |
       +--+--+--+--+--+--+--+--+--+--+--+--+--+--+--+--+

       +--+--+--+--+--+--+--+--+--+--+--+--+--+--+--+--+
    40 |              3            |              F            |
       +--+--+--+--+--+--+--+--+--+--+--+--+--+--+--+--+
    42 |              O            |              O            |
       +--+--+--+--+--+--+--+--+--+--+--+--+--+--+--+--+
    44 | 1  1|                      20                         |
       +--+--+--+--+--+--+--+--+--+--+--+--+--+--+--+--+

       +--+--+--+--+--+--+--+--+--+--+--+--+--+--+--+--+
    64 | 1  1|                      26                         |
       +--+--+--+--+--+--+--+--+--+--+--+--+--+--+--+--+

       +--+--+--+--+--+--+--+--+--+--+--+--+--+--+--+--+
    92 |              0            |                           |
       +--+--+--+--+--+--+--+--+--+--+--+--+--+--+--+--+
```

The domain name for F.ISI.ARPA is shown at offset 20. The domain name FOO.F.ISI.ARPA is shown at offset 40; this definition uses a pointer to concatenate a label for FOO to the previously defined F.ISI.ARPA. The domain name ARPA is defined at offset 64 using a pointer to the ARPA component of the name F.ISI.ARPA at 20; note that this pointer relies on ARPA being the last label in the string at 20. The root domain name is defined by a single octet of zeros at 92; the root domain name has no labels.

4.2 Transport

The DNS assumes that messages will be transmitted as datagrams or in a byte stream carried by a virtual circuit. While virtual circuits can be used for any DNS activity, datagrams are preferred for queries due to their lower overhead and better performance. Zone refresh activities must use virtual circuits because of the need for reliable transfer.

The Internet supports name server access using TCP [RFC-793] on server port 53 (decimal) as well as datagram access using UDP [RFC-768] on UDP port 53 (decimal).

4.2.1 UDP usage

Messages sent using UDP user server port 53 (decimal).

Messages carried by UDP are restricted to 512 bytes (not counting the IP or UDP headers). Longer messages are truncated and the TC bit is set in the header.

UDP is not acceptable for zone transfers, but is the recommended method for standard queries in the Internet. Queries sent using UDP may be lost, and hence a retransmission strategy is required. Queries or their responses may be reordered by the network, or by processing in name servers, so resolvers should not depend on them being returned in order.

The optimal UDP retransmission policy will vary with performance of the Internet and the needs of the client, but the following are recommended:

- The client should try other servers and server addresses before repeating a query to a specific address of a server.
- The retransmission interval should be based on prior statistics if possible. Too aggressive retransmission can easily slow responses for the community at large. Depending on how well connected the client is to its expected servers, the minimum retransmission interval should be 2-5 seconds.

More suggestions on server selection and retransmission policy can be found in the resolver section of this memo.

4.2.2 TCP usage

Messages sent over TCP connections use server port 53 (decimal). The message is prefixed with a two byte length field which gives the message length, excluding the two byte length field. This length field allows the low-level processing to assemble a complete message before beginning to parse it.

5. Master Files

Master files are text files that contain RRs in text form. Since the contents of a zone can be expressed in the form of a list of RRs a master file is most often used to define a zone, though it can be used to list a cache's contents. Hence, this section first discusses the format of RRs in a master file, and then the special considerations when a master file is used to create a zone in some name server.

5.1 Format

The format of these files is a sequence of entries. Entries are predominantly line-oriented, though parentheses can be used to continue a list of items across a line boundary, and text literals can contain CRLF within the text. Any combination of tabs and spaces act as a delimiter between the separate items that make up an entry. The end of any line in the master file can end with a comment. The comment starts with a ";" (semicolon).

The following entries are defined:

```
<blank>[<comment>]

$ORIGIN <domain-name> [<comment>]

$INCLUDE <file-name> [<domain-name>] [<comment>]

<domain-name><rr> [<comment>]

<blank><rr> [<comment>]
```

Blank lines, with or without comments, are allowed anywhere in the file.

Two control entries are defined: $ORIGIN and $INCLUDE. $ORIGIN is followed by a domain name, and resets the current origin for relative domain names to the stated name. $INCLUDE inserts the named file into the current file, and may optionally specify a domain name that sets the relative domain name origin for the included file. $INCLUDE may also have a comment. Note that a $INCLUDE entry never changes the relative origin of the parent file, regardless of changes to the relative origin made within the included file.

The last two forms represent RRs. If an entry for an RR begins with a blank, then the RR is assumed to be owned by the last stated owner. If an RR entry begins with a <domain-name>, then the owner name is reset.

<rr> contents take one of the following forms:

```
[<TTL>] [<class>] <type> <RDATA>

[<class>] [<TTL>] <type> <RDATA>
```

The RR begins with optional TTL and class fields, followed by a type and RDATA field appropriate to the type and class. Class and type use the standard mnemonics, TTL is a decimal integer. Omitted class and TTL values are default to the last explicitly stated values. Since type and class mnemonics are disjoint, the parse is unique. (Note that this order is different from the order used in examples and the order used in the actual RRs; the given order allows easier parsing and defaulting.)

<domain-name>s make up a large share of the data in the master file. The labels in the domain name are expressed as character strings and separated by dots. Quoting conventions allow arbitrary characters to be stored in domain names. Domain names that end in a dot are called absolute, and are taken as complete. Domain names which do not end in a dot are called relative; the actual domain name is the concatenation of the relative part with an origin specified in a $ORIGIN, $INCLUDE, or as an argument to the master file loading routine. A relative name is an error when no origin is available.

<character-string> is expressed in one or two ways: as a contiguous set of characters without interior spaces, or as a string beginning with a " and ending with a ". Inside a " delimited string any character can occur, except for a " itself, which must be quoted using \ (back slash).

Because these files are text files several special encodings are necessary to allow arbitrary data to be loaded. In particular:

of the root.

@

A free standing @ is used to denote the current origin.

\X

where X is any character other than a digit (0-9), is used to quote that character so that its special meaning does not apply. For example, "\." can be used to place a dot character in a label.

\DDD

where each D is a digit is the octet corresponding to the decimal number described by DDD. The resulting octet is assumed to be text and is not checked for special meaning.

( )

Parentheses are used to group data that crosses a line boundary. In effect, line terminations are not recognized within parentheses.

;

Semicolon is used to start a comment; the remainder of the line is ignored.

5.2 Use of master files to define zones

When a master file is used to load a zone, the operation should be suppressed if any errors are encountered in the master file. The rationale for this is that a single error can have widespread consequences. For example, suppose that the RRs defining a delegation have syntax errors; then the server will return authoritative name errors for all names in the subzone (except in the case where the subzone is also present on the server).

Several other validity checks that should be performed in addition to insuring that the file is syntactically correct:

1. All RRs in the file should have the same class.

2. Exactly one SOA RR should be present at the top of the zone.

3. If delegations are present and glue information is required, it should be present.

4. Information present outside of the authoritative nodes in the zone should be glue information, rather than the result of an origin or similar error.

6. Name Server Implementation

6.1 Architecture

The optimal structure for the name server will depend on the host operating system and whether the name server is integrated with resolver operations, either by supporting recursive service, or by sharing its database with a resolver. This section discusses implementation considerations for a name server which shares a database with a resolver, but most of these concerns are present in any name server.

6.1.1 Control

A name server must employ multiple concurrent activities, whether they are implemented as separate tasks in the host's OS or multiplexing inside a single name server program. It is simply not acceptable for a name server to block the service of UDP requests while it waits for TCP data for refreshing or query activities. Similarly, a name server should not attempt to provide recursive service without processing such requests in parallel, though it may choose to serialize requests from a single client, or to regard identical requests from the same client as duplicates. A name server should not substantially delay requests while it reloads a zone from master files or while it incorporates a newly refreshed zone into its database.

6.1.2 Database

While name server implementations are free to use any internal data structures they choose, the suggested structure consists of three major parts:

- A "catalog" data structure which lists the zones available to this server, and a "pointer" to the zone data structure. The main purpose of this structure is to find the nearest ancestor zone, if any, for arriving standard queries.
- Separate data structures for each of the zones held by the name server.
- A data structure for cached data. (or perhaps separate caches for different classes)

All of these data structures can be implemented an identical tree structure format, with different data chained off the nodes in different parts: in the catalog the data is pointers to zones, while in the zone and cache data structures, the data will be RRs. In designing the tree framework the designer should recognize that query processing will need to traverse the tree using case-insensitive label comparisons; and that in real data, a few nodes have a very high branching factor (100-1000 or more), but the vast majority have a very low branching factor (0-1).

One way to solve the case problem is to store the labels for each node in two pieces: a standardized-case representation of the label where all ASCII characters are in a single case, together with a bit mask that denotes which characters are actually of a different case. The branching factor diversity can be handled using a simple linked list for a node until the branching factor exceeds some threshold, and transitioning to a hash structure after the threshold is exceeded. In any case, hash structures used to store tree sections must insure that hash functions and procedures preserve the casing conventions of the DNS.

The use of separate structures for the different parts of the database is motivated by several factors:

- The catalog structure can be an almost static structure that need change only when the system administrator changes the zones supported by the server. This structure can also be used to store parameters used to control refreshing activities.
- The individual data structures for zones allow a zone to be replaced simply by changing a pointer in the catalog. Zone refresh operations can build a new structure and, when complete, splice it into the database via a simple pointer replacement. It

is very important that when a zone is refreshed, queries should not use old and new data simultaneously.

- With the proper search procedures, authoritative data in zones will always "hide", and hence take precedence over, cached data.

- Errors in zone definitions that cause overlapping zones, etc., may cause erroneous responses to queries, but problem determination is simplified, and the contents of one "bad" zone can't corrupt another.

- Since the cache is most frequently updated, it is most vulnerable to corruption during system restarts. It can also become full of expired RR data. In either case, it can easily be discarded without disturbing zone data.

A major aspect of database design is selecting a structure which allows the name server to deal with crashes of the name server's host. State information which a name server should save across system crashes includes the catalog structure (including the state of refreshing for each zone) and the zone data itself.

6.1.3 Time

Both the TTL data for RRs and the timing data for refreshing activities depends on 32 bit timers in units of seconds. Inside the database, refresh timers and TTLs for cached data conceptually "count down", while data in the zone stays with constant TTLs.

6.2 Standard query processing

The major algorithm for standard query processing is presented in [RFC-1034].

When processing queries with QCLASS=*, or some other QCLASS which matches multiple classes, the response should never be authoritative unless the server can guarantee that the response covers all classes.

When composing a response, RRs which are to be inserted in the additional section, but duplicate RRs in the answer or authority sections, may be omitted from the additional section.

When a response is so long that truncation is required, the truncation should start at the end of the response and work forward in the datagram. Thus if there is any data for the authority section, the answer section is guaranteed to be unique.

The MINIMUM value in the SOA should be used to set a floor on the TTL of data distributed from a zone. This floor function should be done when the data is copied into a response. This will allow future dynamic update protocols to change the SOA MINIMUM field without ambiguous semantics.

6.3 Zone refresh and reload processing

In spite of a server's best efforts, it may be unable to load zone data from a master file due to syntax errors, etc., or be unable to refresh a zone within the its expiration parameter. In this case, the name server should answer queries as if it were not supposed to possess the zone.

If a master is sending a zone out via AXFR, and a new version is created during the transfer, the master should continue to send the old version if possible. In any case, it should never send part of one version and part of another. If completion is not possible, the master should reset the connection on which the zone transfer is taking place.

6.4 Inverse queries (Optional)

Inverse queries are an optional part of the DNS. Name servers are not required to support any form of inverse queries. If a name server receives an inverse query that it does not support, it returns an error response with the "Not Implemented" error set in the header. While inverse query support is optional, all name servers must be at least able to return the error response.

6.4.1 The contents of inverse queries and responses

Inverse queries reverse the mappings performed by standard query operations; while a standard query maps a domain name to a resource, an inverse query maps a resource to a domain name.

Inverse queries take the form of a single RR in the answer section of the message, with an empty question section. The owner name of the query RR and its TTL are not significant. The response carries questions in the question section which identify all names possessing the query RR WHICH THE NAME SERVER KNOWS. Since no name server knows about all of the domain name space, the response can never be assumed to be complete. Thus inverse queries are primarily useful for database management and debugging activities. Inverse queries are NOT an acceptable method of mapping host addresses to host names; use the IN-ADDR.ARPA domain instead.

Where possible, name servers should provide case-insensitive comparisons for inverse queries. However, this cannot be guaranteed because name servers may possess RRs that contain character strings but the name server does not know that the data is character.

When a name server processes an inverse query, it either returns:

1. zero, one, or multiple domain names for the specified resource as QNAMEs in the question section
2. an error code indicating that the name server doesn't support inverse mapping of the specified resource type.

When the response to an inverse query contains one or more QNAMEs, the owner name and TTL of the RR in the answer section which defines the inverse query is modified to exactly match an RR found at the first QNAME.

RRs returned in the inverse queries cannot be cached using the same mechanism as is used for the replies to standard queries. One reason for this is that a name might have multiple RRs of the same type, and only one would appear.

6.4.2 Inverse query and response example

The overall structure of an inverse query for retrieving the domain name that corresponds to Internet address 10.1.0.52 is shown below:

```
           +-----------------------------------+
Header     |          OPCODE=IQUERY, ID=997    |
           +-----------------------------------+
Question   |               <empty>             |
           +-----------------------------------+
Answer     |        <anyname> A IN 10.1.0.52   |
           +-----------------------------------+
Authority  |               <empty>             |
           +-----------------------------------+
Additional |               <empty>             |
           +-----------------------------------+
```

This query asks for a question whose answer is the Internet style address 10.1.0.52. Since the owner name is not known, any domain name can be used as a placeholder (and is ignored). A single octet of zero, signifying the root, is usually used because it minimizes the length of the message. The TTL of the RR is not significant. The response to this query might be:

```
           +------------------------------------+
Header     |          OPCODE=RESPONSE, ID=997   |
           +------------------------------------+
Question   |QTYPE=A, QCLASS=IN, QNAME=VENERA.ISI.EDU |
           +------------------------------------+
Answer     |   VENERA.ISI.EDU  A IN 10.1.0.52   |
           +------------------------------------+
Authority  |               <empty>              |
           +------------------------------------+
Additional |               <empty>              |
           +------------------------------------+
```

Note that the QTYPE in a response to an inverse query is the same as the TYPE field in the answer section of the inverse query. Responses to inverse queries may contain multiple questions when the inverse is not unique. If the question section in the response is not empty, then the RR in the answer section is modified to correspond to be an exact copy of an RR at the first QNAME.

CPSIA information can be obtained
at www.ICGtesting.com
Printed in the USA
BVHW01s1811041018
529154BV00063B/540/P